AUSTRALIA'S *Amazing* KANGAROOS

Their Conservation, Unique Biology and Coexistence with Humans

KEN RICHARDSON, MURDOCH UNIVERSITY

National Library of Australia
Cataloguing-in-Publication entry

Richardson, Ken.

Australia's amazing kangaroos : their conservation, unique biology and coexistence with humans / by Ken Richardson.

9780643097391 (pbk.)
9780643097407 (epdf)
9780643107151 (epub)

Includes bibliographical references and index.

Kangaroos.
Kangaroos – Australia.

599.222

Published by
CSIRO PUBLISHING
36 Gardiner Road, Clayton VIC 3168
Private Bag 10, Clayton South VIC 3169
Australia

Telephone:	[+613] 9545 8555
Local call:	1300 788 000 (Australia only)
Fax:	+61 3 9662 7555
Email:	csiropublishing@csiro.au
Web site:	www.publishing.csiro.au

Cover image by iStockphoto
Photographs are by the author unless otherwise stated

Set in Adobe Palatino 10/14 and Trajan Pro
Edited by Bruce Gillespie
Cover and text design by Andrew Weatherill
Typeset by Andrew Weatherill
Index by Russell Brooks
Printed by Ingram Lightning Source

CSIRO PUBLISHING publishes and distributes scientific, technical and health science books and journals from Australia to a worldwide audience and conducts these activities autonomously from the research activities of the Commonwealth Scientific and Industrial Research Organisation (CSIRO). The views expressed in this publication are those of the author(s) and do not necessarily represent those of, and should not be attributed to, the publisher or CSIRO.

Jan26_RP_ILS

Contents

Preface

This book presents a synthesis of many aspects of the biology of the suborder Macropodiformes. Throughout the book I refer to all members of the suborder Macropodiformes as kangaroos. Technically the suborder has three extant families these being the Hypsiprymnodontidae, the Potoroidae and the Macropodidae, and where necessary I will refer to these as Hypsiprymnodonts, Potoroids and Macropodids. This book uses common names as listed in the *CSIRO List of Australian Vertebrates*; however, occasionally I use common names that are more familiar to the audience such as common wallaroo as well as euro.

I have endeavoured to provide an informed basis to areas of debate such as the sustainable harvesting and farming of the larger species. Likewise I try to clarify and explain simply the variety of measures being adopted to resurrect populations of critically endangered, endangered and vulnerable macropodiform species. Like any book of this nature it relies on the enormous energy and endeavours of a great number of past researchers and scholars who, in all their various ways, have built up the scaffolding for current studies of kangaroos. Although this book is not fastidiously referenced, access to the primary literature can be sought using references from the Bibliography. To develop the distribution maps I have used numerous sources to determine distributions, including the IUCN Red Book and data from members of relevant state and territory departments responsible for wildlife under their jurisdiction.

Hopefully this book will bridge the ever-widening gap between the mountains of detailed information found in the serious scientific literature and the many members of the public who wish to be better informed about Australia's iconic kangaroos. In today's rapidly changing world, the better we are informed about our native animals the better their prospects for survival.

ACKNOWLEDGEMENTS

Many people in many walks of life have assisted me over the past 10 years with the gestation of this book. My deep thanks go to: Dr Adrian Bradley, The University of Queensland; Professor Mike Calver, Murdoch University; Dr Jim Cummins, Murdoch University; Dr Robyn Delaney, Northern Territory Government; Dr Andrew Dennis, zoologist, Atherton; Dr Mark Eldridge, Australian Museum; Dr Don Fletcher, Australian Capital Territory Government; Tom Garrett, Macropod & Wild Game Harvesters Assoc Inc.; Professor Colin Groves, Australian National University; Joe Hong, Murdoch University; Professor Emerti Ian Hume, University of Sydney; Peter Johnson, Queensland Government; Dr Deb Kelly, South Australian Government; John Kelly, Kangaroo Industries Association of Australia; Professor Roger Lentle, Massey University; Bob Litchfield, photographer, Ellenbrook; Charlie Manolis, Wildlife Management International Pty Ltd, Darwin; Dr Peter Mawson, Western Australian Government; Samara McPhedran; Dr Keith Morris, Western Australian Government; Dr Diana Nottle, Murdoch University; David Packer, Packer Leather, Narangba; Julia Playford, Queensland Government; Dr David Pearson, Western Australian Government; Dr Lisa Pope; Dr Euan Ritchie, Deakin University; Professor Gordon Sanson, Monash University; Dr Peter Spencer, Murdoch University; and Dr Simon Stirrat, Queensland Government. National Parks and Reserves authorities in all Australian states and the Northern Territory have assisted me in my endeavours to learn about different kangaroo species and their varied habitats. Similarly several postgraduate students and field researchers have also given of their time and knowledge to assist me in my quest to understand the biology of these magnificent animals.

My grateful thanks go to the people who have kindly assisted me with the images and illustrations that make this book visually pleasing as well as helping explain aspects of the stunning animal's biology discussed in this book. I would like to thank the photographers whose extraordinary skills have allowed this book to be so admirably illustrated: Bill Belson, Hans and Judy Beste, Stan Breeden, Ryan Collins (WWF), Brett Dennis, Geoff Griffiths, Martin Harvey, Joe Hong, Jiri Lochman, Marie Lochman, Stuart Miller, Stuart Roper, Dennis Sarson, Raoul Slater, Len Stewart and Dave Watts. I am particularly indebted to Jiri and Marie Lochman for their enthusiasm and support in bringing together the majority of images used in this book. Dr Perdita Phillips, Perth, constructed the distribution maps as well as the beautiful skull and feet illustrations. Dr Natalie Warburton donated her delightful line drawing of kangaroo foot bones.

I am particularly grateful to: Dr Deb Kelly; John Kelly; Dr Peter Mawson; Professor Johanna Plendl, Freie Universitat Berlin; Dr Allan Sheridan, Australian Government; Dr Jeff Short, Wildlife Research and Management Pty Ltd, Kalamunda; and Dr Dominique Thiriet, James Cook University, who each reviewed and made pertinent suggestions to improve draft chapters.

I also wish to express my thanks to my employer Murdoch University, which has and continues to support scholarship and research. Finally, I thank my family especially my children Simon, Rachel and David, who have always been supportive of my endeavours to understand nature.

1. Species characteristics and biology

Introduction

At the beginning of the twenty-first century there are 71 recognised species of kangaroo (Infraclass Marsupialia, Order Diprotodontia, Suborder Macropodiformes) found in Australasia. Of these, 46 are endemic to Australia, 21 are endemic to the island of New Guinea, and 4 species are found in both regions. The various species have a number of common names, including bettong, kangaroo, pademelon, potoroo, quokka, rat kangaroo, rock wallaby, tree kangaroo, wallaby and wallaroo.

Since the advent of the European settlement of Australia in 1788 the endemic kangaroo species have been affected in several ways. Most of the larger kangaroo species – those having a body weight over 20 kg – have had significant increases in their population numbers, as well as expansions of their former distributions. This group includes species with which most Australians are familiar, such as the red kangaroo *Macropus rufus*, the eastern grey kangaroo *M. giganteus*, the western grey kangaroo *M. fuliginosus* and the common wallaroo/euro *M. robustus*. At differing times, all of these have been or are still being harvested commercially.

A second group of kangaroo species has maintained its overall numbers. These species have either a mid-range body weight such as the agile wallaby *M. agilis* that has an average body weight of 11 kg in females and 19 kg in males, or species occupying reliable habitats such as the Herbert's rock wallaby *Petrogale herberti* and the allied rock wallaby *P. assimilis*. All of these species are common and have stable populations.

However, the third group, a large number of species mostly weighing less than 5 kg, have been affected negatively, and have had severe reductions in their numbers as well as in suitable available habitats. The reasons for their decline are many, including altered land use, predation, prevailing fire regimes, and competition from introduced herbivores.

While most of us can readily identify a kangaroo as a kangaroo, we are less able to describe the characteristics we use to determine whether or not an animal is a kangaroo. In most cases its body form, coupled with a bipedal hopping gait, gives us our main visual clues. The characteristic shape of the head, the relatively small thorax with small forelimbs, and the large abdomen with a large tail and hindlimbs, readily identify most kangaroo species. However, it is a series of internal body part characteristics that are conclusive indicators of identity. These range from the plumbing of kangaroos' reproductive tracts to the possession of a sigmoidal tubular stomach, which indicates that the animal is both a marsupial and a kangaroo.

However, in many cases it is not a live animal that we wish to identify, and in some cases not even a recently dead one. In such instances features of the skeleton, especially of the head, are critical to an identification.

The origins of modern kangaroos are unclear, but fossils, as well as modern molecular, immunological and biochemical analyses are gradually advancing our knowledge of their ancestry. Although there are conflicting theories on the basal interrelationships and systematics of the suborder Macropodiformes, in the

Figure 1.1: Western grey kangaroos *Macropus fuliginosus* just north of Perth, Western Australia. (Photo: Jiri Lochman.)

palaeontologic and molecular literature there is broad agreement on most of the detailed measurements of and pertinent characteristic features used to generate the current hypotheses of their evolutionary history.

Recent research strongly suggests that marsupials evolved in the Mesozoic era on the supercontinent Laurasia, probably in what is now known as China, or less likely, in North America. Palaeontological evidence suggests that the last common ancestor of the marsupials and eutherians was early in the Cretaceous period, about 148 million years before the present (mybp) in the northern supercontinent of Laurasia.

A recently discovered small arboreal fossil, *Sinodelphys szalayi*, from geological strata in northern China dated at 125 mybp, has the teeth, ankle and wrist morphologies of a marsupial. At that time, Australia lay far south of where it is today and was still connected to Antarctica and South America, forming the supercontinent Gondwana.

Over the period between 75 and 65 mybp the ancestral marsupials colonised South America, where they evolved into the superorder Ameridelphia, consisting of the order Didelphimorphia (93 extant species) and order Paucituberculata (6 extant species). Molecular evidence suggests that at some time between 70 and 55 mybp small ancestral insectivorous/omnivorous didelphid marsupials spread (possibly twice) from South America west across Gondwana into the Australian continental region. Here they evolved into the superorder Australodelphia that has orders Diprotodontia (137 extant herbivorous species including the phalangerid possums and the Macropodoidea), Dasyuromorphia (71 extant species of carnivore), Peramelomorphia (24 extant species of bandicoot) and Notoryctemorphia (2 extant species of marsupial mole).

In Australia, the oldest fossil marsupials are about 55 million years old, and were collected at Murgon in south-east Queensland. Many of these fossils have strong resemblances to a similar marsupial species (*Dromiciops gliroides*) found in 55-million-year-old rocks in Peru, South America. However, it is unlikely that details of the early marsupial divergences will ever be known, because there are few Australian mammal-bearing fossil deposits covering the period between 55 and 26 mybp.

Molecular evidence suggests that the stem ancestral kangaroo separated from the phalangerid lineage of diprotodonts some time in the Eocene epoch (56–40 mybp). It is thought that an offshoot of these arboreal quadrupedal herbivores reverted to a terrestrial lifestyle, and gave rise to the earliest of the kangaroos, 'protomacropods'. It is most likely that about 45 mybp the early terrestrial 'protomacropods' gave rise to a side branch, which consisted of the ancestral hypsiprymnodontids and the propleopines that retained many primitive features, including a quadrupedal locomotor pattern.

Between 45 and 38 mybp the Australian tectonic plate, carrying Australia, together with what is now the island of New Guinea, separated from Antarctica and moved northwards towards the equator. The immediate consequence of this was that the cold Southern Ocean currents flowed between the two continents, causing massive amounts of ice to form over Antarctica, resulting in Australia becoming much drier.

Over subsequent millions of years, Australia's climate oscillated from being hot and wet (greenhouse conditions with high floral and faunal diversity) to cold and dry (icehouse conditions with a much lower floral and faunal diversity). By about 30 mybp, the island obstructions to the southern oceanic currents running between South America and Antarctica had gone. Consequently the currents flowed freely through Drake Passage and were no longer connected with warmer currents from the lower latitudes. The circumpolar ocean current's temperature dropped about 10°C, making the Australian landmass even colder.

While few Australian mammalian fossils exist for the period 54 to 27 mybp, there are many sites in Australia with rich fossil deposits covering the period 26 mybp (late Oligocene epoch) through to the present time. By the late Oligocene the earliest bipedal hoppers, the bulungamayines, paleopotoroines, potoroines and balbarines, were all present. During the early to middle Miocene epoch (23.5 to approximately 16.4 mybp) Australia had an extended warm and wet period when the eastern coastal region, especially the north, was covered in rainforest. Over this period huge river systems and vast lakes with accompanying luxuriant vegetation dominated the Australian inland. This supported a wide range of marsupial species that occupied the great diversity of available habitats. The hypsiprymnodonts and potoroines were common in the extensive rainforest environments of the time, but there were few macropodine species. Over that period, selective browsing was their principal mode of food acquisition.

About 15 mybp, the north-easterly edge of the Australian Plate, New Guinea, collided with the Pacific Plate. The resultant uplifting of the central spine of New Guinea to heights greater than 4000 metres put Australia in a rain shadow. Australia's annual rainfall declined, average daily temperatures dropped, and the continent gradually slipped into an icehouse period that is still with us today. Floristic changes were profound along the eastern seaboard, and most notably along the north–south-oriented Great Dividing Range (average height 1200 m), where altitude variation, fertile soils, and in the north, warmer climate, all moderated the effects of the continent becoming increasingly cooler and drier. Consequently a variety of rainforest and woodland assemblages remained associated with the Great Dividing Range.

To the west of the range, most of the country is relatively flat, with an elevation averaging only 300 to 450 m. Here climate change, coupled with most of Australia's soils being infertile, resulted in the vast inland rainforests being replaced gradually by sclerophyllous forests and grasslands. As extensive grasslands gradually established over inland areas, grazing became a major factor directing the evolution of new kangaroo species. By about 10 mybp, the balbarines and bulungamayans had disappeared. Between 10 and 5 mybp, as the vegetation became less nutritious, there was a major radiation of the grazing macropodines, with selection pressure for larger and larger individuals. This is because, while the total daily dietary intake to support a large herbivore is high, the amount of food per unit body mass is lower when compared with the more highly metabolically active smaller species. Gigantism of many mainland herbivorous marsupial species resulted.

From about 10 mybp through to the present day there has been significant speciation among the macropodidae. However, many taxonomic interrelationships, as well as the chronology of significant evolutionary events, such as when different genera evolved, are unclear. It is believed that about 10 mybp the New Guinea forest wallaby genera, *Dorcopsis* and *Dorcopsulus*, diverged from the main macropodine lineage.

Both morphological and molecular evidence indicate that tree kangaroos *Dendrolagus* and rock wallabies *Petrogale* are closely related. They are believed to have separated from a common ancestor, possibly one similar to the Proserpine rock wallaby, about 7 mybp. The rock wallabies, which are also closely related to pademelons *Thylogale*, appear to have split from each other about 7.5 mybp. Fossils of both genera occur in the Pliocene epoch (5.3–1.8 mybp) deposits at several sites from north to southern Australia. Ancestral forms of both the tree kangaroos and pademelons are believed to have crossed from mainland Australia to New Guinea by an interconnecting land bridge about 2 mybp.

The largest of the extant macropodines, the red kangaroo, is believed to have been the most recently evolved. It is thought to have evolved in the Pleistocene epoch (2 to 1 mybp).

In more recent times, about 46 000 years ago, there was a mass extinction of the Australian megafauna, including the giant short-faced kangaroo *Procoptodon goliah*, the giant ancestor of today's eastern grey kangaroo *Macropus titan*, as well as of a variety of other very large short-faced kangaroo *Simosthenurus* and wallaby *Protemnodon* species. What caused this is open to speculation.

One possibility is that a rapid cooling and drying of their environment, causing degradation of the available food sources and resulting starvation, brought about their demise. Another possible explanation is that the larger species had slower reproductive rates than their smaller counterparts. An increase in death rates coupled with low reproduction may have tipped populations into a free fall.

Professor Tim Flannery, an eminent palaeontologist, has argued that Aboriginal hunting or, indirectly, their extensive use of patchwork burning, causing significant environmental changes, could have been significant in tipping the balance towards extinction. Other than human involvement, environmental factors such as prolonged droughts have also been suggested as possible significant factors. Following the loss of the Australian megafauna, there was a trend towards smaller body size among the herbivores. The recent fossil history confirms that the extant large kangaroo species evolved from even larger ancestral forms.

At the peak of the last glacial event 18 000 to 20 000 years ago the Australian mainland was connected to both Tasmania and New Guinea, and animal movements between them would have been unimpeded. Over the next 10 000 years, with less and less ice deposition and more and more rain, the sea level gradually rose about 100 m, separating Australia from its two mountainous island neighbours. At about 8000 years ago, many parts of lowland Australia also became isolated from the mainland. Islands such as Barrow, Dorre and Bernier off the mid west coast of Western Australia formed and became refugia for many marsupials, including several macropodine species.

After decades of dedicated work by many palaeontologists and volunteer assistants at sites across Australasia, a large number of fossil marsupial specimens have been found. The subsequent analysis of these specimens has built up a wealth of data that has allowed the core steps in the evolution of macropodids to be elucidated. It is apparent that either two or three monophyletic families form the suborder Macropodiformes.

The first family lineage was that containing the quadrupedal Propleopinae and Hypsiprymnodontinae. Of these, only one hypsiprymnodont is extant today.

The second family, the Potoroidae, had three subfamilies: Palaeopotoroinae, Bulungamayinae and the Potoroinae. Only the Potoroinae are extant today.

The third family lineage, the Macropodoidae, consisted of the Macropodinae, Sthenurinae and Balbarinae. Of these, the Macropodinae and Sthenurinae are represented today.

However, as stated earlier, there are alternate views of kangaroo taxonomy held by different groups of palaeontologists. For instance, Professor Tim Flannery recognised only two families. He included the Propleopinae and Hypsiprymnodontinae in the

Figure 1.2: Adult western grey kangaroo skull. The small cranial vault and narrow snout are typical of macropodids. The medial inflection of the mandible is a characteristic of marsupials. The protuberant pair of lower incisor teeth bite medially onto the upper three pairs of incisors during grazing. (Photo: Joe Hong.)

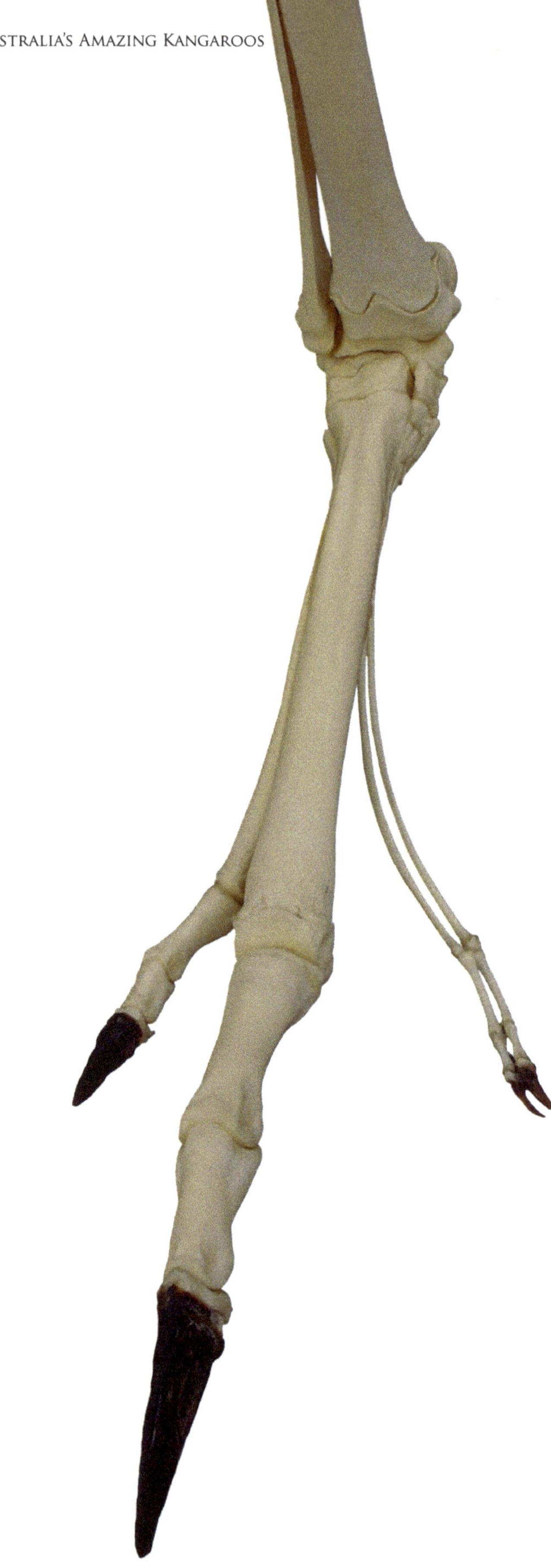

Figure 1.3: A dorsal view of the foot skeleton of a western grey kangaroo showing its large fourth toe (responsible for the scientific name *Macropus*). Note its slender syndactylous toes. (Photo: Joe Hong.)

family Potoroidae. Some palaeontologists believe that the Bulungamayinae are more closely related to the Macropodidae than to the Potoroidae, and should be classified as such. There most likely will always be some degree of speculation about the evolution of the kangaroo!

What features characterise species within the suborder Macropodiformes? All kangaroos are members of the order Diprotodontia, which also contains the closely related possums, koala and wombats. Diprotodonty refers to the characteristics and relationships of the anterior teeth of these animals. Their lower jaw has only a single pair of incisors that are greatly enlarged, chisel shaped and aligned along the horizontal axis of the jaw, so that they are procumbent.

Palaeontologists such as Professor Mike Archer, a doyen in kangaroo palaeontology, use a large number of critical structures that, in sum, define whether or not a specimen is from a kangaroo. They have identified a series of characteristics, such as medial inflection of the angular process of the mandible, a large masseteric foramen, a large masseteric canal, development of the masseteric process, reduction in canine teeth, the loss of the first premolar teeth, enlarged and finely ridged second premolars, a sectorial third premolar, loss of palatal fenestrations, the reduction of the fifth digit of the foot and the presence of epipubic bones, to support their determinations of the major family groups. These allow palaeontologists to key specimens to genus and often to species.

Kangaroos are also characterised by syndactyly of digits 2 and 3 of the hind feet.

Although there has been, and continues to be, healthy debate on details of the taxonomy of the Macropodiformes, there are three recognised extant families: Hypsiprymnodontidae, Potoroidae and Macropodidae. The Hypsiprymnodontidae is a monophyletic group quite distinct from other modern macropodids. The Potoroidae (8 species) have their masseteric canal extending forward to beneath the lower third premolar teeth; both upper and lower third premolars have distinct vertical serrations on their outer surface. There is minimal contact between the squamosal and frontal bones, and there is a proximoventral process on metatarsal 5.

The Macropodidae (41 Australian species), have a relatively short masseteric canal, molars are lophodont, and molar size increases from molar 1 to molar 5.

The taxonomy of Australian modern day kangaroos

Suborder Macropodiformes

Family Hypsiprymnodontidae

Hypsiprymnodon moschatus **Ramsay, 1876**
musky rat kangaroo

Family Potoroidae

Aepyprymnus rufescens **(Gray, 1837)**
rufous bettong

Bettongia gaimardi **(Desmarest, 1822)**
Tasmanian bettong

Bettongia lesueur **(Quoy & Gaimard, 1824)**
burrowing bettong

Bettongia penicillata **Gray, 1837**
brush-tailed bettong

Bettongia tropica **Wakefield, 1967**
northern bettong

Potorous gilbertii **(Gould, 1841)**
Gilbert's potoroo

Potorous longipes **Seebeck & Johnston, 1980**
long-footed potoroo

Potorous tridactylus **(Kerr, 1792)**
long-nosed potoroo

Family Macropodidae

Subfamily Sthenurinae

Lagostrophus fasciatus **(Péron & Lesueur, 1807)**
banded hare wallaby

Subfamily Macropodinae

Dendrolagus bennettianus **De Vis, 1887**
Bennett's tree kangaroo

Dendrolagus lumholtzi **Collett, 1884**
Lumholtz's tree kangaroo

Lagorchestes conspicillatus **Gould, 1842**
spectacled hare wallaby

Lagorchestes hirsutus **Gould, 1844**
rufous hare wallaby

Macropus agilis **(Gould, 1842)**
agile wallaby

Macropus antilopinus **(Gould, 1842)**
antilopine wallaroo

Macropus bernardus **Rothschild, 1904**
black wallaroo

Macropus dorsalis **(Gray, 1837)**
black-striped wallaby

Macropus eugenii **(Desmarest, 1817)**
tammar wallaby

Macropus fuliginosus **(Desmarest, 1817)**
western grey kangaroo

Macropus giganteus **Shaw, 1790**
eastern grey kangaroo

Macropus irma **(Jourdan, 1837)**
western brush wallaby

Macropus parma **Waterhouse, 1845**
parma wallaby

Macropus parryi **Bennett, 1835**
whiptail wallaby

Macropus robustus **Gould, 1841**
euro, common wallaroo

Macropus rufogriseus **(Desmarest, 1817)**
red-necked wallaby

Macropus rufus **(Desmarest, 1822)**
red kangaroo

Onychogalea fraenata **(Gould, 1841)**
bridled nailtail wallaby

Onychogalea unguifera **(Gould, 1841)**
northern nailtail wallaby

Petrogale assimilis **Ramsay, 1877**
allied rock wallaby

Petrogale brachyotis **(Gould, 1841)**
short-eared rock wallaby

Petrogale burbidgei **Kitchener & Sanson, 1978**
monjon

Petrogale coenensis **Eldridge & Close, 1992**
Cape York rock wallaby

Petrogale concinna **Gould, 1842**
nabarlek

Petrogale godmani **Thomas, 1923**
Godman's rock wallaby

Petrogale herberti **Thomas, 1926**
Herbert's rock wallaby

Petrogale inornata **Gould, 1842**
unadorned rock wallaby

Petrogale lateralis **Gould, 1842**
black-footed rock wallaby

Petrogale mareeba **Eldridge & Close, 1992**
Mareeba rock wallaby

Petrogale penicillata **(Gray, 1825)**
brush-tailed rock wallaby

Petrogale persephone **Maynes, 1982**
Proserpine rock wallaby

Petrogale purpureicollis **Le Souef, 1924**
purple-necked rock wallaby

Petrogale rothschildi **Thomas, 1904**
Rothschild's rock wallaby

Petrogale sharmani **Eldridge & Close, 1992**
Mount Claro rock wallaby

Petrogale xanthopus **Gray, 1855**
yellow-footed rock wallaby

Setonix brachyurus **(Quoy & Gaimard, 1830)**
quokka

Thylogale billardierii **(Desmarest, 1822)**
Tasmanian pademelon

Thylogale stigmatica **(Gould, 1860)**
red-legged pademelon

Thylogale thetis **(Lesson, 1837)**
red-necked pademelon

Wallabia bicolor **(Desmarest, 1804)**
swamp wallaby

Recently extinct kangaroos

Potorous platyops **(Gould, 1844)**
broad-faced potoroo

Lagorchestes leporides **(Gould, 1841)**
eastern hare wallaby

Onychogalea lunata **(Gould, 1841)**
crescent nailtail wallaby

Lagorchestes asomatus **Finlayson, 1943**
central hare wallaby

Macropus greyi **Waterhouse 1845**
toolache wallaby

Caloprymnus campestris **(Gould, 1843)**
desert rat kangaroo

Since the European settlement of Australia in 1788, these small macropodid species were last recorded or seen by reliable witnesses at the following times: broad-faced potoroo in 1875; eastern hare wallaby in 1891; crescent nailtail wallaby in the 1930s; central hare wallaby in the 1960s; toolache wallaby in 1972; and desert rat kangaroo in 1974.

The recently described Nullarbor dwarf bettong *Bettongia pusilla* (McNamara, 1997), is known only from recent subfossil material, but it is believed that it may have been extant at the time of European settlement.

The taxonomy of kangaroos of New Guinea and adjacent islands

Suborder Macropodiformes

Family Potoroidae

Aepyprymus rufescens **(Gray, 1837)**
rufous bettong

Family Macropodidae

Subfamily Macropodinae

Dendrolagus dorianus **Ramsay, 1883**
Doria's tree kangaroo

D. d. stellarum **Flannery and Seri, 1990**
Seri's tree kangaroo

Dendrolagus goodfellowi **Thomas, 1908**
Goodfellow's tree kangaroo

D. g. pulcherrimus **Flannery, 1993**
golden-mantled tree kangaroo

Dendrolagus inustus **Müller, 1840**
grizzled tree kangaroo

Dendrolagus matschiei **Forster and Rothschild, 1907**
Huon tree kangaroo

Dendrolagus mbaiso **Flannery, Boeadi & Szalay, 1995**
dingiso

Dendrolagus scottae **Flannery and Seri, 1990**
tenkile

Dendrolagus spadix **Troughton and Le Souef, 1936**
lowlands tree kangaroo

Dendrolagus ursinus **(Muller 1845)**
ursine tree kangaroo

Dorcopsulus macleayi **(Mikluoho-Maclay, 1885)**
Macleay's dorcopsis

Dorcopsulus vanheurni **(Thomas, 1922)**
lesser forest wallaby

Dorcopsis atrata **(Van Deusen, 1957)**
black forest wallaby

Dorcopsis hageni **Heller, 1897**
white-striped dorcopsis

Dorcopsis luctuosa **(D'Albertis, 1874)**
grey dorcopsis

Dorcopsis muelleri **(Schlegel, 1866)**
common forest wallaby

Dorcopsis veterum **Lesson 1827**
grey scrub wallaby

Lagorchestes conspicillatus **Gould, 1842**
spectacled hare wallaby

Macropus agilis **(Gould, 1842)**
agile wallaby

Thylogale browni **(Ramsay, 1887)**
New Guinea pademelon

Thylogale brunii **(Schreber, 1778)**
dusky pademelon

Thylogale calabyi **(Flannery, 1992)**
Calaby's pademelon

Thylogale lanatus **(Thomas, 1922)**
mountain pademelon

Thylogale stigmatica **(Gould, 1860)**
red-legged pademelon

In 2008 a new macropodid species *Dorcopsulus* (sp. nov.) was discovered in Irian Jaya by K. Helgen and has yet to be described taxonomically.

Current status of Australian kangaroos

The internationally recognised organisation for the determination of the conservation status of plants and animals, the World Conservation Union, uses a series of strict criteria to determine the level of risk of extinction that a species faces (see chapter 3). This authoritative organisation makes its determinations, and records these in the International Union for the Conservation of Nature Red List of Threatened Species (IUCN 2011). Of the 50 species of the suborder Macropodiformes, found in Australia, they currently categorise two species as being critically endangered, another five species as being endangered, two species as being vulnerable, 11 species as near threatened, one probable near threatened species as data deficient and the remainder being common to abundant.

The only member of the primitive family Hypsiprymnodontidae, the musky rat kangaroo, is considered to be locally common and at low risk. Within the family Potoroidae, Gilbert's potoroo and the brush-tailed bettong are both listed as critically endangered. Gilbert's potoroo is Australia's rarest mammal. The northern bettong and the long-footed potoroo are both endangered. The burrowing bettong and Tasmanian bettong are both near threatened. The rufous bettong and long-nosed potoroo are in the low risk category.

Within the family Macropodidae, the bridled nailtail wallaby, banded hare wallaby and the Proserpine rock wallaby are endangered. The rufous hare wallaby and quokka are vulnerable. Bennett's tree kangaroo, black wallaroo, parma wallaby, monjon, Cape York rock wallaby, black-footed rock wallaby, brush-tailed rock wallaby, Mount Claro rock wallaby and yellow-footed rock wallaby are all near threatened. The nabarlek is data deficient. Lumholtz's tree kangaroo, western brush wallaby and spectacled hare wallaby are uncommon but are at low risk.

The antilopine wallaroo, black-striped wallaby, tammar wallaby, whiptail wallaby, red-necked wallaby, northern nailtail wallaby, allied rock wallaby, short-eared rock wallaby, Godman's rock wallaby, Herbert's rock wallaby, unadorned rock wallaby, Mareeba rock wallaby, purple-necked rock wallaby, Rothschild's rock wallaby, Tasmanian pademelon, red-necked pademelon, red-legged pademelon and the swamp wallaby are all common over part, or even most, of their range. In some cases, a species may be declining in numbers locally or their historic range may be contracting.

The agile wallaby, western grey kangaroo, eastern grey kangaroo, common wallaroo and red kangaroo are all extremely common.

The Australian species

Family Hypsiprymnodontidae

Musky rat kangaroo
Hypsiprymnodon moschatus

Musky rat kangaroos, commonly called 'stinkers', are the only living members of the family Hypsiprymnodontidae. They are unusual, primitive kangaroos restricted to a narrow band of North Queensland rainforest between latitudes 15°55″ and 18°39″. It appears that the genus Hypsiprymnodon has close affinities with the phalangerid possums as well as with its sister taxon the Potoroidae. It is believed that the Hypsiprymnodon lineage diverged from the ancestral macropod line about 45 mybp.

It is most likely that the ancestral macropodids were frugivores and browsers. With climatic changes, notably a pronounced continental drying, selection favoured the grazers in the drier areas on the one hand and browsers/frugivores in the wetter areas on the other hand. Hypsiprymnodons remained in the wet forests and remained a true frugivore/omnivore.

Musky rat kangaroos are characterised by a unique suite of features, including:

- forelimb and hindlimb proportions that are similar
- five toes on the hind foot; that is, the retention of the first phalange of the first digit, an opposable toe. The retention of the big toe (hallux) is presumed to be a reflection of the arboreal ancestors of the macropodids: the phalangerids. No other extant kangaroo species has this feature
- a prehensile tail that is devoid of hair
- possession of two pairs of lower incisors
- a simple saccular stomach.

Figure 1.4: A musky rat kangaroo foraging during the day in the rainforest undergrowth of north Queensland. These are the only true diurnal members of the suborder Macropodiformes. (Photo: Dave Watts.)

Other characteristics include:

- being diurnal
- not having embryonic diapause
- usually giving births to twins
- most locomotion is by a bounding quadrupedal gait; they do not hop.

They are the smallest of the kangaroos; their body weight ranges from 360 to 680 g for males and 453 to 635 g for females. Measurements of head-body length range from 153 to 273 mm and tail length from 123 to 159 mm. Their overall body colour is a dark brown, with a slightly greyish head and a dark brown scaly tail.

Their skull has a number of specific characteristics, notably two pairs of lower incisors where the first incisor is long and sharp but the second is virtually a stub overlapping the base of the first. Their permanent premolars are narrow, long and have sharply serrated occlusal edges. Each premolar has seven distinct vertical ridges. They have a single lacrymal foramen compared to two in potoroids. Their dental formula is $2\ (I^3_2\ C^1_0\ P^1_1\ M^4_4) = 32$.

Figure 1.5: A musky rat kangaroo. Note that the proportions of its front and back legs are similar. (Photo: Dave Watts.)

These diurnal inhabitants of the north Queensland tropical rainforests can be seen as they quietly forage on the forest floor during daylight hours from sea level (Mission Beach) through to 1500 m (Lake Barrine on the Atherton Tablelands). They have an intimate three-dimensional knowledge of their home ranges, especially of the forest floor.

They are mostly solitary animals but do aggregate around abundant food sources. They are primarily frugivorous, so are dependent on fleshy fruits of forest trees such as figs, quandongs and palms. They assist the ongoing viability of forests with the passage of seeds through their gut, a process that not only increases the probability of the seeds germinating but also ensures their dispersal. They cache seeds by burying them and, when these are not reclaimed and used for food, their germination is facilitated. Invertebrates are taken opportunistically as food.

Musky rat kangaroos have small home ranges that vary from 1 to 4 hectares. They build small nests of leaves, ferns and lichens, which they place adjacent to the base of trees or in low thickets. Reproductive maturity is reached in their second year. They have the shortest recorded gestation period of any kangaroo (19 days). Both postpartum oestrus and embryonic diapause are unlikely. They are seasonal breeders, with between one and three young being born between October and April. Usually two young are raised; however, in 2002 a female road kill was found to have four advanced pouch young. Pouch-dependent young remain in the pouch for about 147 days. Subsequently they leave the pouch, and for a short period are left in the mother's nest, to which she returns regularly to suckle them.

Threatening factors

- Predation by native animals such as pythons, quolls, goshawks and owls.
- Climate changes altering rainforest habitat characteristics.

Management action

- Maintenance of extensive areas of natural habitat. Musky rat kangaroos are unable to live sustainably in small fragmented forest habitats.

Family Potoroidae

It is probable that the family Potoroidae diverged from the Macropodidae in the mid to late Oligocene, about 30 mybp. The Potoroidae consists of five bettong species and three species of potoroo, all of which are small; most adults weigh less than 3.5 kg. All have semi-prehensile tails that have the capacity to curl downwards and hold a bunch of grass or similar vegetation. Potoroids have more strongly clawed front paws and proportionately shorter hind feet than members of the family Macropodidae.

They move slowly on all fours, but use bipedal hopping for speed. Bettongs have a short broad head compared with potoroos, which are more gracile with long slender heads. Their dental formula is $2 (I^3_1\ C^1_0\ P^1_1\ M^4_4) = 30$. Potoroids have a characteristic incisor dentition where their protuberant central maxillary incisors overhang their mandibular counterparts. The lateral and corner maxillary incisors are much smaller. Upper canines are present but small and peg-like. The permanent premolars are large, broad and low with a sharp serrated cutting edge.

Potoroos have 3 to 5 distinct wide vertical ridges on each premolar, while bettongs have 7 to 11 distinct narrow vertical ridges. Behind each premolar lie four low crowned quadritubercular molars that decrease in size from front to back. The occlusal surface of the molars has numerous rounded cusps. The squamosal bone having contact with the frontal bone characterises their skull. They have two lacrymal foramina compared to a single foramen in the hypsiprymnodonts. The auditory bullae in the potoroos are small and flat, but in the bettongs they are large and inflated. All potoroids depend on microbial fermentation of their diet, and as such have a highly modified saccular forestomach where fermentation occurs.

Generally the home ranges of potoroids occupying densely vegetated and wetter forests are smaller than those from drier, less densely forested regions. This is because there is more food available in the wetter areas than in the drier areas. Consequently potoroids in drier habitats forage over large areas to obtain enough food. Prominent in their diet are the underground fruiting parts of fungi, referred to scientifically as hypogeous mycorrhizal sporocarps, commonly called 'truffles'.

Historically, potoroids have played a significant role in turning the soil over as they dig for food. This promotes a better soil organic matter profile, as well as improving drainage. The constriction of their distribution, coupled with lower population numbers of many potoroid species, has lessened their soil-improving effects to the detriment of local environments. The availability of shelter also affects home range size. Bettongs are found in drier grasslands and wooded areas, while potoroos are found in wetter denser environments

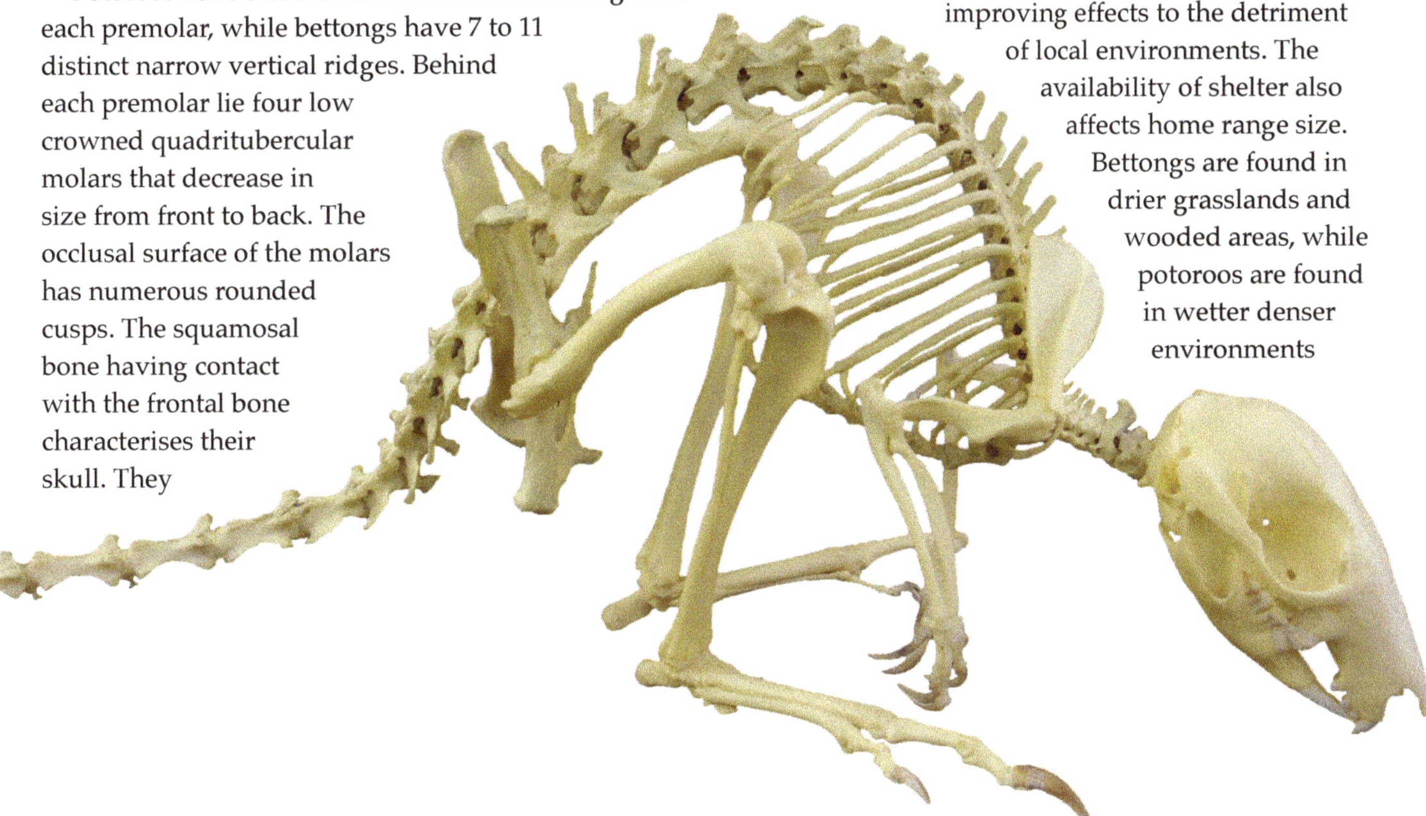

Figure 1.6: Skeleton of a brush-tailed bettong *Bettongia penicillata*. Like all potoroids it has a diprotodont skull, syndactylous hindlimbs, epipubic bones and a long tail. The tail vertebrae have ventrally placed chevron bones that run the length of the tail. These act as insertion points for tendons and facillitate the ventral prehension that these bettongs display when carrying nest material. (Photo: Joe Hong.)

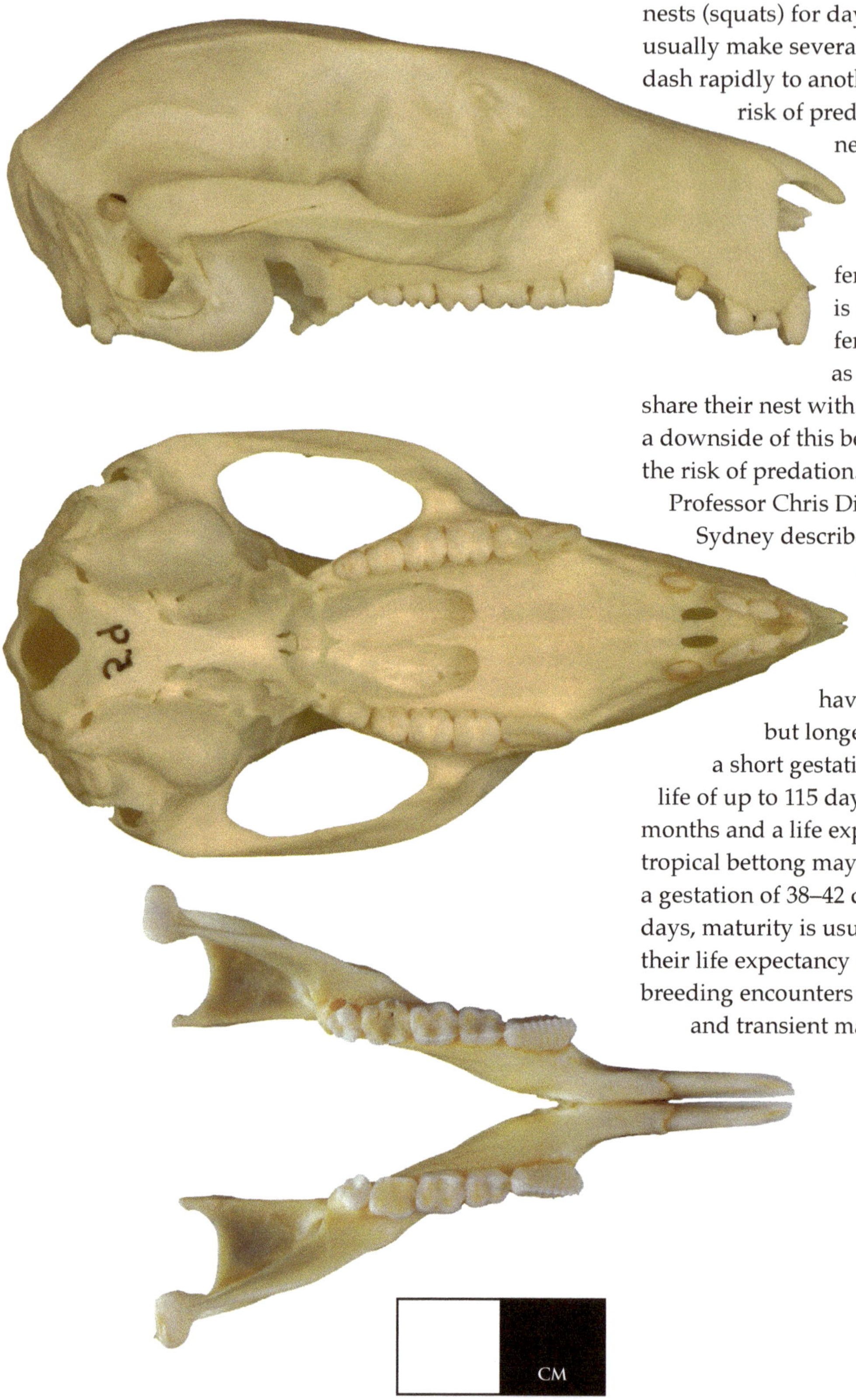

such as heathlands and forests having dense undercover.

Potoroos and bettongs use well-camouflaged nests (squats) for daytime predator avoidance. They usually make several nests and, if disturbed in one, dash rapidly to another nearby, thus lessening the risk of predation. They often abandon a nest from where they have been disturbed. While nest sharing by adults is uncommon, it does occur when males check the female's reproductive status. It is possible that at such times the female assesses the male's 'fitness' as a potential mate. Females usually share their nest with their young-at-foot. However, a downside of this behaviour could be an increase in the risk of predation.

Professor Chris Dickman from the University of Sydney describes two life strategies among the potoroines: the bettongs are fast track breeders with a short lifespan, compared with the potoroos, which have a slower breeding strategy but longer lifespan. The bettongs have a short gestation period, 21–24 days, a pouch life of up to 115 days, maturity between 7 to 10 months and a life expectancy of 4 years. However, the tropical bettong may live much longer. Potoroos have a gestation of 38–42 days, a pouch life of up to 150 days, maturity is usually between 1 and 2 years and their life expectancy is up to 7 years. Opportunistic breeding encounters occur between oestrous females and transient males.

CM

Figure 1.7: Brush-tail bettong skull. Note the large auditory bullae, the fine vertical grooves on the premolars and the peg-like canines. (Photo: Joe Hong.)

Bettongs

Aepyprymnus rufescens **(Gray, 1837)**
rufous bettong

Bettongia gaimardi **(Desmarest, 1822)**
Tasmanian bettong

Bettongia lesueur **(Quoy & Gaimard, 1824)**
burrowing bettong

Bettongia penicillata **Gray, 1837**
brush-tailed bettong

Bettongia tropica **Wakefield, 1967**
northern bettong

Rufous bettong *Aepyprymnus rufescens*

Throughout its extensive range it is commonly called the 'rufous rat kangaroo'. This heavily built species, the largest of the bettongs, is found over a wide geographic area. It occurs in a wide band from Cooktown in far north Queensland to the northern rivers area of coastal New South Wales in the south. A disjunct population is found in the Barrington Tops region of NSW. Recently the species was found in New Guinea.

Adult body weights range from 1.3 to 3.5 kg. Measurements of head-body length range from 345–400 mm in males to 363–480 mm in females. Tail length averages 344 mm in males and 357 mm in females. Ear length ranges from 48 to 57 mm. They have a characteristic shaggy, light-coloured, rufous coat that has emergent silvery guard hairs. They have a white belly and a distinctive, lightly coloured distal tip to the tail. They adopt a characteristic upright stance when alert.

They frequent many habitats on both sides of the Great Dividing Range, ranging from sclerophyllous forests to dry open woodlands to coastal eucalypt forests. Rufous bettongs prefer habitats with a relatively open understorey, usually of native grasses such as *Poa* spp. and *Imperata* spp. They are locally common, particularly in the coastal Spotted Gum *Eucalyptus maculata*-dominated forests having gently sloping aspects. Their home ranges vary from 75 to 110 hectares (ha) for males and 45 to 60 ha for females. Individual animals 'rest up' during the day in one of several nests (they may have up to five in use at any one time) that they have made within their territory. Their globular grass nests are placed on a shallow scrape usually abutting a grass clump, log or the base of a bush or tree. Sometimes they use burrows to escape the extremes of summer heat. They are strictly nocturnal.

Their diet depends on the seasonal availability of browse, herbs and grasses. In some circumstances grasses can be up to 80 per cent of their diet. Their well-developed forepaw claws allow them to dig efficiently and effectively to obtain significant amounts of roots, tubers and insect larvae. Truffles (fungal sporocarps) are not common in their diet because these are uncommon in the dry grassy woodlands. Rufous bettongs have been observed to scavenge carcasses, a practice that has been seen in several other kangaroo species. Under extreme circumstances individuals

Figure 1.8: Rufous bettong, startled during the day by the photographer. (Photo: Jiri Lochman.)

Figure 1.9: A rufous bettong foraging for roots and tubers. (Photo: Jiri Lochman.)

may move up to 4.5 km when foraging. Under most environmental conditions they obtain enough water from their diet; consequently they rarely drink water.

Small polygamous associations (one male with two females) are the most common grouping, but solitary animals are common. The sex ratio is near parity. Sexual maturity is at about 12 months for males and about 9 months for females. In favourable years breeding can occur throughout most of the year, resulting in three young being born and weaned, but there is a cessation of breeding over harsh summer periods.

Oestrus occurs at 3-week intervals. Because an oestrus female has only a 12-hour receptive period, males improve their possibility of access to these females by remaining close by throughout that period. Gestation is 22 to 24 days. They have embryonic diapause. The females mate within three to four hours of the birth of their single offspring. The newborn young moves immediately into the pouch, where it remains attached to a teat for the next 7 to 8 weeks. Pouch emergence occurs a further 7 to 8 weeks later. The young leaves the pouch permanently at the end of the fourth month but remains with its mother for the next 2 months.

Threatening factors

Currently there are no major threats to this species.

- However, fox predation occurs over most of their current distribution except for small areas of northern Australia.
- Continuing loss of habitat is potentially problematic in the long term, but currently this is not a problem.

Management action

- Measures to limit fox numbers would be advantageous.
- Lessen competition from other herbivores such as rabbits.

Tasmanian bettong *Bettongia gaimardi*

Historically they were called the eastern or southern bettong; however, this usage has diminished now that the original large mainland population has become extinct. This small bettong is now restricted to eastern Tasmania and a few nearby offshore islands (Maria and Bruny).

They weigh between 1.2 and 2.3 kg. Their measurements are: head-body length 315–332 mm, tail length 288–345 mm and ear length 29–40 mm. Their upper body is a pale rufous colour with light grey-white emergent hairs. The underparts are pale to white. The tail generally has the terminal third dark and the tip white.

They are most often found in open mature eucalypt or casuarina forests growing on infertile doleritic soils. These forests have an understorey of short native grasses and/or sedges with some larger shrubs. Because of the soil infertility of their preferred habitats these small bettongs have large home ranges (38–85 ha). Their day nests are large, densely woven, globular structures of grass and bark placed on a scrape on the ground against sheltering structures such as logs and shrubs. Adults do not exhibit strong nest area territoriality but they do have strong site fidelity. The latter characteristic has the potential advantage of improving their knowledge of truffle locations. They know the home range terrain intimately, allowing them to more readily escape predators. Males know the female nest locations and they monitor the reproductive status of females.

Tasmanian bettongs are strictly nocturnal, and may move up to 1.5 km in a night between their nest and feeding grounds. They spend most of their nights foraging for truffles that they locate by odour. The sparce, clumped distribution of truffles appears to be a major factor in the bettongs having large home ranges. They complement their diet with roots, bulbs and seeds. They are believed to eat exudates from acacias.

These bettongs reach sexual maturity between 9 and 12 months. From then on they are continuous breeders. Their gestation period is 21 days and they have an oestrous cycle of 23 days. The young remain in the pouch for about 105 days. They are weaned between 165 and 180 days.

Fire has a negligible effect on mortality. In the absence of foxes, post-fire mortality from predation is low.

Threatening factors

Currently these bettongs are relatively common and numbers remain stable.

- Clear felling of native forests in Tasmania has made large areas unsuitable for bettongs because the uniform regrowth, especially of young stands, is an unsuitable habitat.
- The malicious introduction of foxes into Tasmania with the eventual probable predation of these bettongs is alarming. These bettongs are poorly equipped to respond to novel predators such as foxes. It is believed that their predator naïvity was a major factor in the extermination of the extensive mainland bettong population of the nineteenth and twentieth centuries.
- Inappropriate use of fire.

Management action

- Ensure that foxes are eradicated from Tasmania.
- Modify forest harvesting techniques to leave movement corridors.
- Retain a reasonable proportion of suitable older growth forest.
- Establish and maintain suitable dry sclerophyll forests with open understoreys.
- Manage the use of fire to improve the available food resources especially truffles.

Figure 1.10: A female Tasmanian bettong accompanied by her large young foraging on the forest floor. These bettongs need relatively large areas of mature forests for their survival. (Photo: Dave Watts.)

Burrowing bettong *Bettongia lesueur*

These are commonly called 'boodies'. Before the European colonisation of Australia this species had an extremely widespread distribution through much of Australia's arid and semi-arid zones. Shortly after Europeans colonised Australia, burrowing bettong numbers began to decline because of changes in land practices as well as predation by introduced foxes and feral cats.

In recent times burrowing bettongs have been restricted to Bernier, Dorre, Boodie and Barrow islands off the coast of Western Australia, where they can occur at densities in excess of 12 per ha. Following their accidental introduction onto Boodie Island, black rats *Rattus rattus* caused extreme environmental damage. In 1985, Western Australian Government authorities undertook a poisoning program to eradicate the rats. This was successful, but also eradicated the few remaining burrowing bettongs. Since then burrowing bettongs have been successfully reintroduced back onto Boodie Island, as well as Faure Island and Heirisson Prong in Western Australia, the Arid Recovery Reserve in South Australia and Scotia Sanctuary in New South Wales. On Bernier and Dorre islands their numbers have fluctuated dramatically in the period 2005 to 2010 because of local drought.

Fencing predators out and maintaining predator reduction–eradication programs within these areas have been the basis for the success of these

Figure 1.11: Burrowing bettong foraging among succulents on Dorre Island. (Photo: Jiri Lochman.)

Figure 1.12: Burrowing bettong on Barrow Island. (Photo: Jiri Lochman.)

enterprises. Currently animal numbers on their island refugia, as well as at their reintroduction sites, are stable, and in most cases numbers are increasing.

Their natural habitats on Bernier, Dorre and Barrow islands are characterised by low vegetation. Here, burrowing bettongs are found primarily associated with the dunes and dune edges. Their burrows occur either in the vicinity of limestone capping rock, where successful burrows can be dug and maintained, or along dune edges where the sand is more compacted. Burrows range from simple tunnels with only one or two entrances through to extensive warrens with many entrances. Burrowing bettongs are social animals with, in some instances, up to 40 individuals occupying a single warren. Historically, on the Australian mainland, loamy soils were preferred burrowing sites. Living in burrows during the day lessens their daily water turnover as well as lowering the probability of predation.

Burrowing bettongs weigh between 900–1600 g and have a head-body length of about 360 mm, and tail length of about 280 mm. Animals from Barrow Island are significantly smaller than those from Bernier and Dorre islands. Burrowing bettongs have a creamy to light brown upper pelage, with emergent white and grey hairs. Their under body is paler. The tail is creamy-brown with a black subterminal third that often has a white tip.

Burrowing bettongs are strictly nocturnal. In the evenings they may move up to 5 km from their burrows to their feeding grounds, where roots and tubers are the principal components of their diet. However, browse, forbs, fruit, seeds, fungi and soil arthropods are eaten whenever they are available. Burrowing bettongs have been observed eating carrion.

Males have a territory that may overlap those of between one and six females. Male–female associations appear to be a harem arrangement, a way of life that possibly accounts for them being a highly vocal species. Sexual maturity is reached at about the fifth month. Breeding is seasonal with peaks in April–May, and possibly November–December. They have a summer anoestrus. Their gestation period is 21 days. Pouch reliance is 115 days. The population turnover, sometimes in excess of 25 per cent per year, is believed to be driven by adult behaviour towards juveniles as well as occasionally the effects of drought. Subadult survival equals that of adults.

Threatening factors

Currently, even though bettong numbers fluctuate dramatically with seasonal vagaries, animal numbers on their more northern island refugia appear to be stable. However, these bettongs are highly susceptible to:

- predation by cats and foxes
- fire
- novel disease.

There are limited suitable habitats remaining on the mainland where they could be reintroduced.

Management action

- Maintain predator control in all colonies.
- Continue reintroduction programs to suitable mainland habitats.
- Continue ongoing habitat management and monitoring of all populations.
- Ensure rodents or novel predators are not accidentally introduced onto the bettongs' island refugia.
- Prevent threatening factors such as the introduction of disease to animals on their offshore island refugia.

Figure 1.13: Burrowing bettong burrow entrances at Heirisson Prong, Shark Bay. Their warrens can be very extensive and have numerous entrances. (Photo: Jiri Lochman.)

Brush-tailed bettong *Bettongia penicillata*

Commonly called 'woylie' in south-western Western Australia. This is currently Australia's second most threatened macropod, yet before European colonisation this species was common over large tracts of land south of the tropics, such as the three deserts region (Sandy, Great Sandy and Gibson Deserts) of Western Australia and the area from Eyre and York Peninsulas north to central South Australia, then diagonally across to the western edge of the Great Dividing Range of northern New South Wales.

Currently this species has naturally occurring populations at Dryandra woodland, Tutanning Nature Reserve and Perup Forest in southern Western Australia. The Perup population has two distinct genetic components despite there being no physical or geographic barriers to explain this situation. Their preferred habitats are either open forests or woodlands (wandoo and jarrah), which are characterised by low clumped understorey areas. Other than altered land uses, the introduction of foxes is believed to be the principal cause of the decline in their numbers.

In southern Western Australia, there are large areas with many species of native gastrolobium plants. These plants have high concentrations of sodium fluoroacetate, which is a naturally occurring anti-herbivore metabolite concentrated in their leaves and seeds. This is extremely toxic to mammal species that have not coevolved with it. Western Australian brush-tailed bettongs have coevolved with the gastrolobiums and consequently have high levels of liver enzymes capable of rapidly degrading sodium fluoroacetate.

The generalised resistance of Western Australian marsupials to fluoroacetate has been capitalised upon by Government authorities, who use poison baits containing low levels of fluoroacetate, commonly called 1080, to kill foxes throughout much of the brush-tailed bettong's former range, in the knowledge that the poison will not affect the resident native mammals. As a consequence of the baiting regimen fox numbers declined and until quite recently brush-tailed bettong numbers had increased to high levels in some areas of southern Western Australia.

Their body weight ranges from 0.75–1.85 kg. Their head–body length ranges from 280–380 mm, tail length 250–360 mm, and ear length averages 28 mm. Their overall coat colour is grey-brown with emergent grey and rufous guard hairs. They have characteristic pale eye-rings and they have a tuft of dark hair towards the tip of the tail.

Brush-tailed bettongs' home ranges vary according to habitat, but for males are generally 28–43 ha and for females 15–28 ha. Within the home ranges there are smaller, seasonally changing, feeding ranges (up to 9 ha) that in turn contain core nesting areas of 2–3 ha, the latter being defended by the males but not by the females. Adults have strong site fidelity. Philopatry is greater in females than in males. They are strictly nocturnal; returning to their large grass and bark, dome nests each day. Brush-tailed bettongs are known to remove the outer bark in strips from the base of stringybark trees for their nests. In some areas, such as the jarrah forests of Western Australia, this has resulted in the base of jarrah trees having a characteristic light-coloured area shredded to a height of 30–40 cm. The bark is used to line their nests. Like other bettongs, brush-tailed bettongs can hop with nest materials held in their downwardly curled tail.

They regularly travel long distances from their daytime refugia to their evening foraging sites. Throughout the year, more than 50 per cent of their diet consists of truffles. This is supplemented by tubers, bulbs, seeds, grasses, invertebrates and occasionally hakea gum. Their primary food source, truffles, is closely associated with the roots of specific native trees, especially acacias. Truffles are clumped and patchily distributed over wide areas. Consequently brush-tailed bettongs may travel long distances between patches of high quality forage.

Fire is a common component of their environment. Brush-tailed bettongs will forage in the warm ashes following a bushfire even though there may be hot embers and the loss of cover exposes them to predation. This appears to be because some fungi fruit within an hour of a fire passing over them. Brush-

Figure 1.14: A brush-tailed bettong in a submissive posture to another bettong following an aggressive encounter. (Photo: Jiri Lochman.)

Figure 1.15: Just before dawn a brush-tailed bettong uses its tail to carry nest material at Dryandra. They are capable of carrying large amounts with their downturned tail. (Photo: Jiri Lochman.)

tailed bettongs are also known to consume seeds and, in some instances, cache these. Caches that are not revisited enhance seed germination. The commercially valuable Western Australian Sandalwood *Santalum spicatum* has been advantaged by this activity.

Breeding is continuous. Their oestrus cycle is 22 days and gestation period is 21 days. Pouch life is about 90 days. Young reach maturity quickly, so females are able to wean about three young each year. Offspring survivorship in the wild is between 11 and 15 per cent. In the wild their lifespan is between 4 and 6 years.

Reintroduction into dozens of sites in southern Western Australia since the early 1990s was quite successful as long as effective predator control, particularly of foxes, was maintained. These successes were also a result of the adoption of stringent protocols for the capture, transportation and release of the brush-tailed bettongs. However, in 2001 a rapid decline in the number of brush-tailed bettongs occurred across most of their range. By 2006 some populations had disappeared, others were severely depleted – a 90 per cent decline – while others were less affected. The causes of this dramatic widespread decline had not been determined by 2011.

Although there has been an extraordinary level of research into why the species has had such a dramatic decline, there is no conclusive answer to this simple question. It appears that the declines are density dependent. Hypotheses such as cyclic environmental changes, predation and unknown disease processes have been suggested and examined to no avail. It is probable that the reason is multifactorial. The brush-tailed bettong was listed as critically endangered in 2008 because there had been such a dramatic decline (80–90 per cent) in their population over 10 years.

Threatening factors

- Predation by foxes and cats.
- Climatic change affecting plant community compositions.
- Habitat destruction through land clearances by humans, feral pigs disrupting and destroying the understorey of the woodlands, and damage to plant communities by the 'Dieback' pathogen *Phytophthora cinnamonii*.
- Fire.
- Novel disease.

Management action

- Maintain fox control throughout the areas in which brush-tailed bettongs are currently distributed.
- Develop and implement effective cat control in areas occupied by brush-tailed bettongs.
- Maintain a fire regimen that lessens the probability of severe destruction of the critical understorey of their prime habitats.
- Continue reintroductions of disease-free brush-tailed bettongs into suitable habitats.
- Maintain research into the biology of the species as well as of the causal agent(s) of the widespread decline of the species.
- Ensure that long-term monitoring of populations is in place.
- Ensure that novel disease entities are not introduced to the populations.
- Improve the education of the public about the threats to, and importance and management of, brush-tailed bettongs.

Figure 1.16: The large globular nest of a brush-tailed bettong in Wandoo woodland. (Photo: Jiri Lochman.)

Northern bettong *Bettongia tropica*

Also called the 'tropical bettong'. Northern bettongs are an endangered species that frequent a restricted habitat west of the Great Dividing Range in far north Queensland. Their preferred habitat is a narrow band (less than 10 km wide) of forest between the tropical rainforest edge on the east and the arid zone immediately to the west. The eastern aspect is fire-prone wet sclerophyllous eucalypt forest, and further to the west are drier eucalypt (corymbia) woodlands. The area has a distinct wet season between January and April.

These bettongs are found in three areas: Carbine Tableland, Lamb Range and Coane Range (Paluma); a possible fourth site is the Windsor Tableland. Of these, the Lamb Range population is the healthiest, having reasonable numbers spread over a wide area. The other populations are small and each is restricted to a small area. All of these habitats are extremely tenuous, where the bettongs may have a high risk of possible extinction. Trapping studies have shown that northern bettongs prefer the higher slopes and ridges of the tall moist eucalypt and allocasuarina forests. Within these they prefer areas having high densities of woody shrubs with sparse groundcover and low densities of native grasses.

These small potoroines are monomorphic. Their body weight ranges from 0.9 to 1.4 kg. Their head-body length is 275–432 mm, tail length is 268–350 mm, and ear length is 38–40 mm. They are characterised by a soft grey pelage, gracile limbs and a tail tipped with dark hair.

They are strictly nocturnal. In the daylight hours they shelter in well-hidden grass or bark oval-shaped nests, mostly located at the base of a tree, in the skirt of a grass tree or dug in adjacent to a fallen log. These are primarily for avoidance of predators. Generally

Figure 1.17: A pair of northern bettongs, each carrying nesting material in its downturned prehensile tail. (Photo: Hans and Judy Beste.)

they have several closely spaced nests that allow them to dart from one shelter to the next if they are disturbed. Individuals move relatively long distances from their nests to available food each day. Northern bettongs are restricted in their distribution by the rainforest to the east and by drier soils beyond the rain shadow to the west. They feed extensively on truffles, more than 45 per cent of their diet, depending on location and season. They also forage extensively on cockatoo grass *Alloteropsis semialata* and lilies *Hypoxis* spp.

Seasonal fire is important because it keeps the invading rainforest (mostly from the east) at bay. Photographic evidence suggests that over the past 50 years there has been a major invasion by the rainforest into the dry sclerophyll forest along the length of the Lamb Range. Regular cyclic burning over a timescale of 3 to 5 years is advantageous because this maintains a vigorous growth of kangaroo grass *Themada australis* as well as causing a post-fire increase in truffle formation, notably the fire-adapted *Mesophellia* spp., especially over the first 3 years. This is followed by a senescent decline in sporocarp numbers. After about 5 years kangaroo grass is tall and lank and ready to be burnt again.

Since their food sources are usually dispersed over large areas, the northern bettong's home ranges are large, about 50–70 ha, in both males and females, but

Figure 1.18: Two northern bettongs in their partially built nest at the base of a large forest tree. (Photo: Hans and Judy Beste.)

may reach 120 ha. Truffles have a patchy distribution, and the density of these varies with time, area and species. This results in male bettongs being more likely to pair with a single female and remain monogamous than would be the case if food were readily available and population densities high. The latter, in particular, would encourage more home range overlap and promiscuity. Their lifespan is believed to be between 7 and 12 years.

Females commence breeding at 300 days and males at 440 days of age. Breeding is continuous throughout the year and is asynchronous. Mate guarding is a significant factor in determining paternity and breeding success. However, their breeding potential is quite high. Their oestrous cycle is between 21 and 23 days and gestation period is 20–23 days. Postpartum oestrous occurs within a day, and is followed by embryonic diapause. Pouch life is for 102–112 days. Weaning occurs between 165 and 185 days. They are capable of raising about three offspring to independence each year. Consequently, one would expect their numbers to be reasonably high. It is most likely that offspring survival is a critical factor. In captivity this can be as high as 60 per cent, but in the wild it is probably about 10 per cent.

Population turnover appears to be low. Although adult mortality is low, there is a low level of recruitment (12 per cent) from births and immigration. Because males disperse over longer distances, about 1.3 km compared with 0.9 km in females, they have a greater effect on the gene pool than the females. In her doctoral thesis Lisa Pope suggests that there is limited gene flow in the Lamb Range population even over 9 km. It appears that since the Pleistocene period two genetically separate populations have arisen, and both occur in the Lamb Range area.

Even though there appears to be continuous suitable habitat, behavioural traits have acted as dispersal barriers. Dr Pope argues that genetic structure is determined by past and current dispersal barriers, density fluctuations, dispersal patterns and mating systems. Fires and seasonal changes appear to have little effect on dispersal and recruitment. She concludes that individual populations appear stable with low probabilities of extinction.

According to the 2011 IUCN Red List of Threatened Species, the northern bettong is endangered because its geographic distribution is less than 5000 sq. km, their actual area of occupancy is less than 500 sq. km, their population is fragmented and found at no more than five sites, and there has been a decline in the quality and extent of their habitat.

Threatening factors

- There is limited appropriate habitat. This is closely linked to the seasonal abundance, diversity and availability of their primary food source, truffles.
- Loss of the wet sclerophyll habitat because of rainforest incursion and use by man for farms and houses.
- Altered fire regimens, resulting in senescence of their understorey habitat, causing a drop in the availability of truffles and lowered palatability of kangaroo grass.
- Cattle grazing.
- Competition with feral pigs for food as well as habitat destruction by the pigs.
- Fox and probably cat predation if they become established in their area. Foxes have extended their range into nearby areas of far north Queensland.
- Limited genetic diversity and inbreeding depression.

Management action

- Limit the loss of habitat.
- Maintain a fire regimen that improves food availability as well as reversing the rainforest incursion into prime bettong habitat.
- Reintroduce northern bettongs into suitable sites in their former range.
- Undertake genetic studies to determine the feasibility of restocking smaller populations without causing detriment to the genetic diversity of the species.
- Undertake separate captive breeding programs for both the northern and southern populations.
- Ensure appropriate cattle grazing regimens are implemented.
- Cull the resident pig population.
- Prevent foxes entering the area.
- Ensure that novel disease entities are not introduced into the population.

POTOROOS

Potorous gilbertii **(Gould, 1841)**
Gilbert's potoroo

Potorous longipes **Seebeck & Johnston, 1980**
long-footed potoroo

Potorous tridactylus **(Kerr, 1792)**
long-nosed potoroo

Gilbert's potoroo *Potorous gilbertii*

Gilbert's potoroo is currently Australia's rarest mammal. John Gilbert collected the first recognised Gilbert's potoroo from King George's Sound in the south-west of Western Australia and named it *Hypsiprymnus gilbertii*. However, through the nineteenth and twentieth centuries authorities erroneously identified the potoroos caught in the south-west of Western Australia as either the long-nosed potoroo *Potorous tridactylus* or the now extinct broad-faced potoroo *Potorous platyops*. After not being collected for more than 100 years, Gilbert's potoroo was rediscovered in 1994 on Mt Gardner in Two People's Bay Nature Reserve, near King George Sound where the type specimen was collected. Studies in the 1990s confirmed that *P. gilbertii* is a full species.

Historically they were found only in the high rainfall zone (average annual rainfall 900-plus mm) of southern Western Australia. Currently they only occur naturally in the Mt Gardner area, which is a granitic structure having extremely dense, low *Melaleuca striata* heathlands with interspersed open ground. They are small potoroos that are brown-grey above and paler below. Long fur on the face make their eyes appear to bulge from the face and for the ears to be buried. Their body weights range from 710–1200 g, head-body length measures 280–380 mm, and tail 200–236 mm. They have small front feet with long curved claws.

Field details for this species are limited because of the potoroo's rarity, its sensitive natural habitat and the extremely small area in which it is found. Potoroos have home ranges of up to 15–25 ha in males and 3–6 ha in females. They lie up in small rounded grass nests during the day. They are nocturnal, and travel up to a kilometre to forage for truffles that comprise 90 per cent of their diet throughout most of the year. They have been recorded to consume in excess of 40 species of fungi. However, they are most likely omnivorous, eating roots, fleshy fruits and soil arthropods.

Their reproductive biology is little known. Animals are sexually mature at 12 months. Breeding occurs throughout the year. Their gestation period is less than 38 days. They remain in the pouch for about 4 months. Young remain within their mother's home range for between 7 and 18 months. Their lifespan is greater than 10 years.

Gilbert's potoroo is the most threatened of Australia's kangaroos. According to the 2011 IUCN Red List of Threatened Species, they are critically endangered because their population size is low and estimated to be fewer than 50 mature wild individuals in the Mt Gardner population. Currently their natural occurrence is limited to about 8 sq. km and their area of occupancy is less than 5 sq. km. The Western Australian Department of Environment and Conservation has a reintroduction program by which more than 50 individuals have been successfully bred on nearby Bald Island. Although this is encouraging, their low genetic diversity suggests that their long-term future is bleak. A second captive population has been established in a 360 ha fenced enclosure at Norman's Beach, Waychinicup National Park.

THREATENING FACTORS

Even though numbers are low, they are believed to be stable. Threats to any expansion of the current population are many, including:

- fire: their current natural populations on Mt Gardner live in long-unburnt habitat. Fire to this could be catastrophic
- fox and cat predation
- limited appropriate habitat, hence limitations to the seasonal abundance, diversity and availability of their prime food source, truffles
- loss of habitat because of ongoing land clearances nearby

Figure 1.19: A female Gilbert's potoroo in heathland habitat at Two People's Bay. Note her beautiful long claws. (Photo: Jiri Lochman.)

Figure 1.20: A female Gilbert's potoroo with dependent young at foot. Both are foraging among the leaf litter in heathland. This habitat is highly sensitive to bushfire. (Photo: Jiri Lochman.)

- dieback (*Phytophthora cinnamonii*), a plant fungal disease common throughout most of southern Western Australia, which has had and is having a major effect by killing the Proteaceous plants that form much of the protective overstorey of the potoroo's habitat. A variety of *Phytophthora* species are also believed to have reduced the abundance of truffles, which are the major food item for these potoroos
- the low genetic diversity of the potoroo. Low recruitment of young is a serious impediment to the recovery of the species to sustainable populations. Consequently the population has limited genetic ability to accommodate novel environmental catastrophes. It is most likely that they were never common and have possibly been on the brink of extinction for quite some time.

Management action

- Use all possible measures to protect the small area of suitable habitat that they currently occupy from threatening processes, especially fire and plant diseases.
- Prevent predation by cats and foxes.
- Increase studies into the biology of the species.
- Endeavour to extend the current area of suitable habitat.
- Identify new areas of suitable habitat.
- Ensure that novel disease entities are not introduced to the population.
- Limit human interference with the potoroo population.
- Establish a number of breeding colonies that are self-sustaining.
- Investigate and improve techniques to improve reproductive success.
- Continue the reintroduction program.
- Improve the public education about the Gilbert's potoroo, its importance, threats and management.

Long-footed potoroo *Potorous longipes*

This endangered species was first discovered in 1967 in the East Gippsland region of Victoria. They are large potoroines, with males weighing up to 2.3 kg and females 1.7 kg. Their general body measurements are: head-body length 380–415 mm; tail length 315–325 mm; and hind foot length 103–114 mm. They are grey-brown above and paler below. They are characterised by having a large foot length relative to head length, as well as having an additional footpad, the hallucal pad.

Long-footed potoroos are found at elevations between 100 and 1100 m above sea level. The vegetation of these areas varies, but in general terms they live in moist (average annual rainfall 1100–1200 mm) sclerophyllous forests, temperate rainforests and riparian forests that have a dense understorey. Constant moist soil appears to be critical to the understory plant communities essential to their well-being.

Currently this species is restricted to three separate geographic areas. There are scattered populations in East Gippsland, and a second quite large population is in the Barry Mountains of north-eastern Victoria. The third area is in south-eastern New South Wales near the Victorian border, where numbers are extremely low and their presence is known from predator scats examinations and hair-trap analyses.

These potoroines are strictly nocturnal. During the day they lie up in globular nests well hidden in the dense understory. Their sex ratio is parity. Home ranges vary geographically, with larger home ranges occurring in East Gippsland (22–60 ha) compared to those of the Barry Mountains (14–23 ha). Most long-footed potoroos are monogamous, and female home ranges usually fit within that of their male. Home range size presumably reflects the overall abundance of food. In this case truffles can account for up to 90 per cent of their total dietary intake, the remainder being invertebrate and plant material. Like all potoroids heavily reliant on truffles, they eat the fruiting bodies of a large number of fungal species. This follows the fungal fruiting phenology throughout the year. The seasonal changes in the occurrence of truffles modify how potoroos use their home range. Pairs forage together.

Animals reach sexual maturity at about 2 years of age. They are continuous breeders; however, in some areas more births occur in spring. This is probably a reflection of the greater abundance and better quality of food available in the months before they give birth. Gestation period is about 38 days. Pouch life is 140–150 days. They have the potential to wean between 2.5 and 3 young in a year. They are a long-lived species, often living for more than 8 years in the wild.

According to the 2011 IUCN Red List of Threatened Species, they are endangered because their geographic distribution is less than 5000 sq. km; in this case they occupy about 1830 sq. km. The East Gippsland population occupies about 1120 sq. km, the north-eastern Victorian population about 500 sq. km and the New South Wales population about 200 sq. km. Their population is severely fragmented and occurs at no more than five sites; here there are three widely separated subpopulations. There has been a continuing decline in animal numbers at a number of sites. Overall the entire population is believed to number fewer than 2500. Captured individuals number fewer than 500. The estimate of there being up to 2500 individuals in total is based on extrapolations from potential available suitable habitat and known potoroo biology.

Threatening factors

- Limited appropriate habitat, hence limitations to the seasonal abundance, diversity and availability of their prime food source, truffles.
- Predation by feral dogs, dingoes, foxes and cats.
- Inappropriate fire regimens.
- Habitat disturbances, such as timber harvesting, road building and fuel reduction burning regimens.
- Feral pigs are believed to be competitors for the truffles and also disturb the understorey, lessening the quality of the habitat.

Figure 1.21: An alert long-footed potoroo scanning the undergrowth. The long nails of its forepaws are a characteristic of all potoroids. (Photo: Dave Watts.)

Figure 1.22: Female long-footed potoroo and large dependent young at foot. (Photo: Dave Watts.)

Management action

The Victorian Government carefully manages its national parks to ensure that the long-footed potoroo populations are well protected and are self-sustaining. Where potoroo colonies occur outside of national parks the Victorian Government has in place a number of pertinent management guidelines that include establishing Special Management Areas (SMA) of 400–500 ha where these potoroos occur in areas of state forest.

- No fuel-reduction burning is being undertaken within the SMAs, and where possible the surrounding buffer zone is also left unburnt.
- Control of dingoes, wild dogs, foxes, cats and pigs is being improved.
- More detailed management of logging and road building in and near the SMAs is being undertaken to minimise their effects on known potoroo populations.
- There are attempts to conserve and improve potential new areas for future expansion and/or relocation of long-footed potoroos.
- Long-term monitoring of populations is in place.
- There is a continuing research program on the distribution and ecology of these potoroos.
- Research on food resources, in particular the biology of the hypogeous fungi, is a priority.
- A captive colony has been established to facilitate ongoing basic research, development and improvement of husbandry techniques.
- It is essential to ensure that novel disease entities are not introduced to the populations.
- Public education about the long-footed potoroo, its importance, threats and management is ongoing.

Long-nosed potoroo *Potorous tridactylus*

This is the most widely distributed and abundant of the potoroos. They are found in moist coastal forests and heathlands (annual rainfall of more than 760 mm) from south-east Queensland to north-east Victoria through to Tasmania and King and Flinders islands. It is relatively uncommon throughout its patchy mainland distribution, but is common in Tasmania. In mainland Australia the population consists of numerous, small, scattered subpopulations.

These potoroos are characterised by a proportionately short hind foot and the absence of a hallucal footpad. Their body weight ranges from 660–1640 g. Body measurements are: head-body length 260–410 mm; tail length 200–250 mm; and hind foot length 70–82 mm. Their coat colour varies from a deep brown to grey above and much paler to white below. Across their range of distribution animals have a gradient in the occurrence of a white tail from none in southern Queensland to 80 per cent in Tasmania. When they hop their tail is extended directly backwards and the forepaws are held close to the body.

They appear to be partly diurnal. They rest up in the dense undergrowth for most of the day and emerge to feed in the late afternoon and early evening. They use a complex series of runways through their densely vegetated habitats (heathlands and the understorey of sclerophyllous forests) to access their foraging areas. Individuals rarely move far from dense undergrowth. As with other potoroos, truffles are significant in their diet. They also eat roots and tubers as well as invertebrates, especially insect larvae.

They are solitary sedentary animals. The sex ratio is less than parity, and biased in favour of females. Reported home ranges vary, from a Tasmanian site of 5.2–19.4 ha for males and females respectively to a south-western Victorian site of 2.9–4.0 ha for males and females respectively, reflecting a higher quality habitat in Victoria. Generally the home range of males overlaps those of several females. Both males and females build and maintain several (up to seven) nests within their home range. Simple depressions are dug at the base of large grass clumps, bushes or grass trees where the foliage usually covers the animal. Generally individuals occupy their own nest, but there may be temporary cohabitation by the resident male of a female's nest. Cohabitation of a female with her young-at-foot is common.

Males reach sexual maturity at about 8 months and females at about 12 months. On mainland Australia they are asynchronous continuous breeders but have a period of anoestrus between March and June. In Tasmania they have a late summer and a winter–early spring breeding season. Their gestation period is 38 days and their pouch life between 120 and 130 days. As with most potoroids they are able to wean up to three offspring each year. Following weaning, young males disperse widely and take up residence in any available suitable habitat. This mechanism allows them to most efficiently occupy a fragmented and temporally changing habitat found over large areas. Some young females also disperse, but not as widely. Low levels of subadult survival and recruitment result in low density in mainland populations. The healthier populations on Tasmania probably reflect the absence of foxes.

Threatening factors

Mainland populations appear to be declining, while those on Tasmania are stable.

- Predation by cats, feral dogs and foxes.
- Loss of suitable habitat.
- Fire.

Management action

- Limit predation.
- Address the ongoing loss of suitable habitats.
- Change inappropriate fire regimens.

Figure 1.23: A long-nosed potoroo holding its food with its long claws. (Photo: Jiri Lochman.)

Family Macropodidae

This family consists of two extant subfamilies, the Sthenurinae, containing only the banded hare wallaby and the Macropodinae, that includes 40 species currently found in Australia that can be grouped as follows: (a) the 'typical' kangaroos, wallabies and wallaroos of the genus *Macropus* (13 species) as well as the monospecific genus *Wallabia* (the swamp wallaby); (b) rock wallabies *Petrogale* (16 species); tree kangaroos *Dendrolagus* (2 species); pademelons *Thylogale* (3 species); (c) nailtail wallabies *Onychogalea* (2 species); (d) 'typical' hare wallabies *Lagorchestes* (2 species); and (e) the monospecific genus *Setonix* (the quokka).

They are all characterised by a small head, small conical thorax and a large abdomen supported by extremely well-developed hindlimbs and a strong tail. Their forelimbs are small and dextrous, while their feet are large with little lateral or rotational flexibility. The range of movement of the distal hindlimbs is restricted primarily to a hinge-like action. Their feet have a strong, elongate, robust fourth toe that bears the brunt of propulsive and impact forces. Kangaroos use bipedal hopping as their principal mode of locomotion. This is energetically inefficient at low and high speeds, but very efficient at intermediate speeds. Some species use their tail as a balancing organ as well as a lever during slow pentadactyl locomotion.

Except for the tree kangaroos, kangaroos can only move forwards; they cannot move backwards or move their legs independently. Their skull is characterised by the parietal bone contacting the alisphenoid bone.

Kangaroos are nocturnal, and generally 'rest up' during the day in protected sites such as dense thickets of vegetation and rocky outcrops. They move to feeding grounds in the late afternoon and early evening to partake of their herbivorous diet. Most typical kangaroos are grazers, but the tree kangaroos, rock wallabies, pademelons, quokka and swamp wallaby are all browsers. The adaptations to these differing modes of food acquisition and processing lie in dental specialisations, notably of their incisor and cheek teeth. Most macropods return to their daytime resting sites immediately before or just after dawn.

When a female becomes fertile, her pheromonal state encourages the attendance of nearby males. When she is receptive, mating occurs and pregnancy generally results. A very short gestation period of 3 to 5 weeks ensues leading to the birth of a very small precocial neonate. The young, commonly called a 'joey', instinctively uses its well-developed forelimbs to climb from the mother's cloaca through the belly fur to enter the pouch where it attaches firmly to one of the mother's four teats. Over the next few months the joey grows steadily until it is able to relinquish its permanent hold on the teat, explore the pouch and poke its head out of the pouch until eventually it can leave the pouch. Most free-living joeys remain in close contact with and suckle their mothers for several weeks before being weaned.

Figure 1.24: A red-necked wallaby *Macropus rufogriseus* browsing acacia shrubs in Tasmania. (Photo: Jiri Lochman.)

Figure 1.25: A family group of whiptail wallabies *Macropus parryi* in open forest at Carnarvon National Park in south-east Queensland. These gregarious wallabies may form groups of up to 50 individuals. (Photo: Dave Watts.)

SUBFAMILY STHENURINAE

Banded hare wallaby *Lagostrophus fasciatus*

The taxonomy of this unusual small kangaroo is unclear. One point of view is that it is the only living member of the short-faced kangaroos; that is, a member of the subfamily Sthenurinae. Another view is that they belong to a separate subfamily, the Lagostrophinae. They are characterised by having upper and lower incisors that bite together, a feature differing from the other members of the family Macropodidae, where the upper incisors lie lateral to the lower incisors. Sthenurids have a straight and flat molar row. Another unusual feature is their possession of a small premolar 3, that quickly falls out allowing the cheek teeth to migrate slowly forward during the animal's lifetime.

Banded hare wallabies occur naturally only on Dorre and Bernier islands off Shark Bay, where their population was estimated in 2010 to be between 2300 and 1800 individuals. Because they have poor predator avoidance strategies, attempts to reintroduce them back onto mainland Australia have been unsuccessful, where cats and to a lesser extent foxes are present. An attempt to reintroduce them into the wild at Francois Peron National Park in 2001 failed because of intense predation by feral cats. The recent reintroductions onto Faure Island in Shark Bay, which does not have cats or foxes, have been highly successful.

Average measurements of Dorre Island banded hare wallabies are: weight 1.62 kg; head-body length 430 mm; head length 79 mm; tail length 312 mm; crus length 123 mm; and pes length 105 mm. Other than having a slightly smaller pes length, they are statistically identical to Bernier Island animals in all body measurements. They are characterised by their fast, agile locomotion and having a naked muzzle. They have a dense greyish-brown pelage with sparse emergent white hairs and eight to 10 dark transverse bands across their lower back and rump.

During the daytime, animals rest up in thick, shady thickets. On Dorre Island they are particularly fond of consolidated dunes running the length of the island

Figure 1.26: A banded hare wallaby on Dorre Island. Its shaggy coat and distinct banding across the back is characteristic of this unusual short-faced kangaroo. (Photo: Marie Lochman.)

Figure 1.27: A foraging banded hare wallaby on Faure Island in Shark Bay, Western Australia. Their recent reintroduction onto this island has been highly successful. (Photo: Jiri Lochman.)

that are heavily vegetated with *Ficus*, *Diplolaena*, *Acacia* and *Heterodendrum* species. On Bernier Island they are mostly found in *Acacia*/*Heterodendrum* thickets. Banded hare wallabies are crepuscular, and mostly browse available shrubs and graze on grasses and herbs.

Hare wallabies are seasonal breeders, with most pouch young appearing in late summer and early autumn. If they lose their pouch young or their young-at-foot it is unlikely that there will be a blastocyst ready for implantation, so they have to wait until the next season to replace the pouch young. Their gestation period is between 29–31 days. They have a postpartum oestrus. Pouch life is about 6 months, and weaning is 3 months later. Adult females normally rear a single young each year. The adult sex ratio of males:females is 0.8:1. Longevity in the wild is between 5 and 6 years, and up to 12 years in captivity.

According to the 2011 IUCN Red List of Threatened Species, they are endangered because they occur over an area of less than 5000 sq. km in total and their occupancy is less than 500 sq. km. Also, their population is known to exist at no more than five locations, and there are extreme fluctuations in the number of mature individuals.

Threatening factors

Currently the banded hare wallaby is listed as being endangered. The main threatening factors include:

- the introduction of predators such as cats and foxes onto their island refugia
- loss of suitable habitat
- catastrophic extensive fire
- novel disease.

Management action

- Maintain vigilance to prevent the introduction of predators onto their island refugia. Maintain predator control in translocation sites.
- Ensure appropriate informed management of all areas in which populations occur.
- Prevent novel threatening factors, such as disease, to animals on their offshore island refugia.
- Continue reintroduction programs to suitable island and mainland habitats.
- Ensure fires, especially large ones, do not occur in their colonies.

Subfamily Macropodinae

TREE KANGAROOS

Tree kangaroos are believed to have evolved in Australia between 5 and 7 million years ago, probably from a rock wallaby ancestor, possibly an ancestral Proserpine rock wallaby. Given that the baseline kangaroo lineage was derived from an arboreal phalangerid possum, tree kangaroos are secondarily arboreal. They have had to reacquire a suite of specialised features for them to be able to move efficiently in their arboreal lifestyle.

Tree kangaroos are arboreal folivores and frugivores that currently live mostly in rainforests. They forage selectively a wide variety of plant species (growing shoots, flowers, fruits) but avoid eucalypts, probably because they contain chemicals (especially phenolics and terpenes) toxic to their foregut microbes. Their dental formula is 2 $(I^3_1\ C^1_0\ P^1_1\ M^4_4) = 30$. The auditory bulla is only slightly expanded.

Ancestral tree kangaroos are believed to have moved across an ancient land bridge from Australia to New Guinea between 2 and 4 million years ago where they radiated into a variety of niches. Of the 12 species of extant tree kangaroo, only Bennett's tree kangaroo and Lumholtz's tree kangaroo are found in Australia. It is believed that the two Australian tree kangaroo species are relict populations following the loss of the ancient land bridge. Both of these species are little known because they live in rainforests, often in inaccessible terrain or in isolated areas of north Queensland. Both species are cryptic and secretive in their habits. Allopatry between Lumholtz's and Bennett's tree kangaroos is maintained by interspecific competition.

The other 10 species are found on the island of New Guinea and a few offshore islands in Irian Jaya. In New Guinea they are widespread, and are found mostly in the montane rainforests. Of the New Guinea tree kangaroo species, the grizzled tree kangaroo *Dendrolagus inustus* retains the primitive macropodid feature of having an elongate middle toe. This ancestral feature is only shared with the two Australian tree kangaroo species. These three species are grouped as long-footed tree kangaroos. The other nine tree kangaroo species are 'short-footed', having a shortened middle toe making them better adapted for an arboreal lifestyle.

Historically, ancestral tree kangaroos were widespread west of Australia's Great Dividing Range. Recent studies have described two ancestral tree kangaroo species from middle Pleistocene fossil deposits on the Nullarbor region of south central Australia. This suggests that during that period the region would have been covered with open woodlands and grasslands capable of supporting large arboreal tree kangaroos. This is quite a different scenario to that found in the region today.

To facilitate their arboreal lifestyle, tree kangaroos have large strong forelimbs with extremely long, tough claws. When climbing they use their forelimbs to grasp the tree above their head and then push off with their hindlimbs to lever themselves upwards. To do this they have a number of adaptations to their limb architecture. Compared to their terrestrial relatives they have more equal hindlimb-to-forelimb ratios of muscle and bone. Their forelimbs are longer and more robust than in terrestrial macropodids. They have a mesh of shoulder tendons to hold their shoulder joint tightly together when they reach above their head to pull themselves upwards. In addition their highly mobile wrist joints allow medial grasping of tree limbs, and well-developed forelimb muscles ensure that they can maintain their grip. Because they rarely need to hop quickly there has been a significant lessening in the muscle mass of their hindlimbs. The tree kangaroo's main mode of propulsion is to lever itself up tree trunks and along inclined branches using its hindlimb muscles. The anatomy of their tarsal joint (ankle) allows flexibility for them to grasp their substrate. Their hind feet are short, increasing their ability to lever themselves when climbing. They also have broad, soft, spongy footpads that ensure safe footing and lessen the possibility of falling. Even so, examination of tree kangaroo skeletons shows that many have had serious falls resulting in broken limbs at some stage of their lives. Unlike other macropods they can move their hind legs independently when moving slowly. They have been reported to leap in excess of 9 metres down to lower tree branches and up to 18 m to the ground, without injury. This appears to be because of their lower leg tendons, such as the Achilles, being much thicker than those found in terrestrial kangaroos of comparable size. A thick, long tail assists them with balancing.

Olfaction plays a major role in their social communication. Paracloacal and sternal glands both produce pheromones for social communication, and in particular for maintenance of home ranges. Tree kangaroos are the only known kangaroos that will defend their home range. Because of their arboreal lifestyle and the complexities of obtaining high-quality non-toxic food sources in their forest environments, female tree kangaroos make a significant investment in maternal care. The long association of mother and offspring improves the likelihood of subadults successfully occupying new territories.

Figure 1.28: Lumholtz's tree kangaroo *Dendrolagus lumholtzi* browsing forest leaves. Note that its tail is longer than its body length. (Photo: Dave Watts.)

Bennett's tree kangaroo
Dendrolagus bennettianus

Figure 1.29: A Bennett's tree kangaroo eating a grevillea blossom on a pale moonlight night. (Photo: Martin Harvey.)

This, the larger of Australia's two species of tree kangaroo, is now restricted to roughly 3650 sq. km of lowland rainforest between the Daintree River and Mt Amos immediately south of Cooktown. They are sexually dimorphic, with the body weight of males ranging from 11.5 to 13.7 kg and of females from 6.3 to 10.6 kg. Their head-body length ranges from 540 to 750 mm. Their ears are short and rounded. Their pelage is a nondescript dark brown-grey with a greyish forehead and snout. The neck and upper shoulders are pale rufous. The forearms, paws and feet are black. Compared to other tree kangaroos, they have a relatively long tail, 730 to 840 mm, that is thick for its entire length. It is usually pale for most of its length but is dark towards the tip.

Field data is limited for this species. Bennett's tree kangaroos occupy a variety of habitats ranging from lowland rainforests, to notophyll vine forests through to riverine gallery forests. They mostly sleep during the day, often in sunny places. They are nocturnal browsers of the forest canopy and are highly skilled at manoeuvring through the canopy.

Their diet ranges from leaves – presumably they select the more highly nutritious ones – to fruits, especially of the various native figs *Ficus,* and olives *Chionanthus* and *Olea*. Roger Martin, an ecologist who spent a considerable amount of time studying this species in the forests around Shipton's Flat in the Annan River Valley, reports that individual animals have their own feeding preferences for a small suite of tree, shrub and vine species from the many hundreds of species available.

Analysis of leaves of preferred species, such as native longans *Dimocarpus australianus,* scaly bark ash *Ganophyllim falcatum,* wild randa *Aidia racemosa* and the umbrella tree *Schefflera actinophylla,* as well as a suite of vine species, confirms that all have high nitrogen levels. Younger leaves generally have higher concentrations of crude fibre, nonstructural carbohydrate as well as of phosphorus and potassium. They also have low concentrations of lipid, cell wall carbohydrates, lignin, calcium, iron and selenium.

Males occupy large territories of 20–30 ha that they defend vigorously against intruding males. Females have much smaller home ranges of 6–11 ha. They have a polygynous mating system that is facilitated by the home ranges of a male overlapping those of several females. Females may be seasonal breeders, because most give birth late in the dry or early in the wet season. Their gestation period is 44 days. It is unlikely that embryonic diapause occurs. Pouch life is about 9 months.

Threatening factors

- Loss of suitable habitat.
- Predation by dogs and dingoes, as well as by amethystine pythons *Morelia amethistina,* could become a problem.

Management action

- Maintain the habitat throughout its current limited distribution.
- Use appropriate grazing and fire regimens to manage areas of good habitat.
- Lessen the threats of predation by dogs and dingoes.
- Improve the knowledge base of their biology as well as of the habitats in which they are currently found.

Figure 1.30: A Bennett's tree kangaroo, where one can see the extremely strong sharp claws of its paws. These are essential for it to be able to climb quickly and efficiently. (Photo: Martin Harvey.)

Lumholtz's tree kangaroo
Dendrolagus lumholtzi

It is found in an area of about 5500 sq. km of montane rainforests, mostly above 800 m, between the Cardwell Range and the Mount Carbine Tableland of north Queensland. In this region high quality forests growing on basaltic soils support greater numbers of tree kangaroos than forests found on less fertile metamorphic soils. Consequently they are most common in the drier forests of Herberton, and least common in wetter areas of Lamb Range (Tinaroo/ Davies Creek). Whether this is a result of absolute moisture levels or the leaching of nutrients from the wetter soils has not been determined. Disturbed forests support greater numbers of animals than older established forests because in their regrowth phase they produce more and have a greater diversity of food.

Lumholtz's tree kangaroos are sexually dimorphic. Their body weight averages 8.6 kg for males to 7.05 kg for females. Head-body length is 520 to 710 mm, and tail length ranges from 470 to 800 mm. This species has a dark grey-black upper body including the face, and a paler undersurface of the body. They are characterised by a pale rim of fur that runs from the neck along the back of the face to the top of the head and between the ears. This light-coloured rim contrasts markedly with the dark face. The short, rounded ears are generally pale in colour.

Because of their small body size and climbing technique these kangaroos prefer food trees having trunks that are small-to-moderate size in circumference. These tree kangaroos are nocturnal, sleeping during the day in dense vine thickets or on suitable branches usually between 6 and 13 m above the ground. They often sleep in sunny sites. To escape from arboreal predators they quickly shin down smaller trees devoid of large epiphytes. Once on the ground they hop to safety before climbing back into the trees.

Tree kangaroos are broad generalist and opportunistic folivores, highly skilled at manoeuvring through the canopy. At night they spend much of their time between 6 and 10 metres above the ground browsing on leaves, fruit and flowers. They are especially partial to the fruits of native figs and olives. Because tree kangaroos feed high in the canopy, detailed information of their true diet is difficult to obtain. Observations suggest that they select the more highly nutritious leaves of a wide variety of plant species. Analysis of the leaves they select shows that nitrogen levels are high and that structural carbohydrates are low.

They are a sedentary species, where males have an average home range of about 2 ha, while those of females average 0.7 ha. The home ranges of several females may overlap that of a male. They are often in small groups of three to five individuals, usually a male with several consorts and juvenile animals. Pheromones are used to define and maintain home ranges. Males are highly territorial and vigorously defend their territory against intruding males. An orange-coloured pigmentation (possibly cinnabarinic acid powder) has been observed on the upper thighs and belly of sexually mature male Lumholtz's tree kangaroos. Females move only short distances; consequently in fragmented habitats they may all be closely related.

Sexual maturity is at about 2 years in females and 4.5 years in males. They have a polygynous mating system. They are continuous but slow breeders, with births occurring at any period of the year. Their oestrous cycle is between 47 and 64 days and their gestation period is between 42 and 48 days. A postpartum oestrous and embryonic diapause probably does not occur. The sex ratio of pouch young is parity. The joey stays in the pouch for between 35 and 39 weeks. It first exits between 26 and 33 weeks, and weaning is generally between 30 and 60 days after final pouch exit. Young stay with their mother for up to 2 years. It is believed that this is to educate the young on food choices, because the nutritional quality and palatability of leaves varies with their stage of growth and the time of year. Also, this period could possibly be when the young learn to avoid toxic plant products that also vary seasonally.

Figure 1.31: Lumholtz's tree kangaroo resting on a rainforest tree limb. (Photo: Stan Breeden.)

Figure 1.32: A juvenile Lumholtz's tree kangaroo sitting on a bird's nest fern. The mantle of lighter coloured hair that characteristically delineates their face in adults has not developed at this stage. (Photo: Hans and Judy Beste.)

Orphaned tree kangaroo pouch young in captivity need to have an innoculum of their mother's gastric contents to ensure their well-being. Presumably this is so that they can develop an appropriate microbial flora to facilitate their transition from a reliance on mother's milk to that of plant fermentation.

Threatening factors

- They are vulnerable to predation by pythons, dogs and dingoes.
- Animals in forests recently affected by land clearances and logging frequently lose body condition and even starve to death.
- They are particularly vulnerable when crossing roads, and are commonly found as road kill.

Management action

- Wherever possible, current unused small plots of natural rainforest on freehold and government land should be held *in perpetuity* as habitat for tree kangaroos.
- Ongoing revegetation, of suitable areas not viable for agriculture, is needed to provide corridors between strategic blocks of preserved forest fragments.
- Walkways from adjacent forest canopies over roads has proven effective in lessening tree kangaroo road kills; more need to be put in place.
- Long-term control of feral dogs and dingoes is needed to limit predation.

True hare wallabies

These are small, arid–semi-arid zone wallabies that have rapid graceful movements. Early settlers believed that their fast acrobatic movements were similar to those of the European hare. At the turn of the twentieth century they were common across much of Australia's arid–semi-arid zone. Now, in the early twenty-first century, to Australia's shame the central hare wallaby and the eastern hare wallaby have become extinct.

Of the two remaining species, the spectacled hare wallaby, even though its numbers have declined across much of its former range, is still relatively common. However, the rufous hare wallaby now only occurs naturally on Bernier and Dorre islands off the mid-west coast of Western Australia, and has been successfully reintroduced to Trimouille Island in the Montebello Island group, off the Western Australian Pilbara coast.

Figure 1.33: Rufous hare wallaby *Lagochestes hirsutus* foraging in its natural habitat on Bernier Island. (Photo: Jiri Lochman.)

Spectacled hare wallaby
Lagorchestes conspicillatus

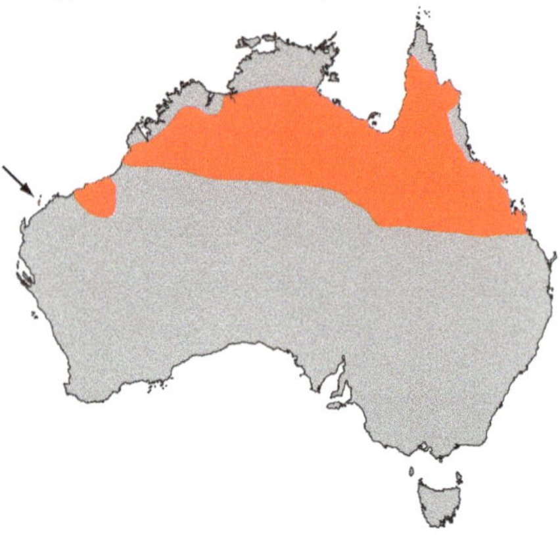

This is a large hare wallaby (1.6–4.7 kg) that is characterised by startling orange eye-patches and white fur around its muzzle. Its back and upper flanks are a dark grey-brown, with emergent grey hairs, giving an overall grizzled appearance. Its underparts are a pale grey to white. Head-body length ranges from 340 to 490 mm, and tail length ranges from 370 to 530 mm.

These wallabies are widely distributed across the drier tropical regions of northern Australia. The species is secure across most of its range in the Northern Territory and Queensland, where it is found in tussock (*Triodia*) grasslands that have irregularly distributed shrub thickets. Home ranges may be as large as 177 ha in poorer habitats such as the tussock grasslands of central northern Australia, but elsewhere, in better quality habitats, they can be as small as 8–10 ha.

In Western Australia they have declined in abundance across their northern distribution, and are only secure on Barrow Island. An isolated population immediately south of Broome in the open grazing country has been confirmed recently from several road kills on the Great Northern Highway. In 1997 this species was confirmed present in the Trans-Fly region of New Guinea. The Barrow Island subspecies has recently been reintroduced to Hermite Island in the Montebello group after it was exterminated there some time after 1912 by feral cats.

They are nocturnal animals that lie up in scrapes under large tussocks or similar structures during the day. Some lie up in burrows built at the base of large hummocks to avoid predators as well as high ambient temperatures. Physiologically, spectacled hare wallabies are extremely well adapted to living in dry environments. Their daily water turnover and urine production are both low, indicating that they obtain their water from their food. They are browsers that select the more highly nutritional plant parts of non-chenopod shrubs, forbs and grasses. Seeds, fruits and invertebrates are taken opportunistically. Unusually among the kangaroos, in hard times they can subsist on the tips of spinifex leaves.

Sexual maturity is reached at 12 months in both males and females. They are continuous breeders. They have a postpartum oestrus and embryonic diapause. Their gestation period is between 29 and 31 days. Pouch life is 5 months, allowing them to wean two offspring each year.

Threatening factors

Although the species is considered secure, its numbers are believed to be in decline. Possible reasons are:

- predation by foxes and feral cats
- too frequent burning of the spinifex grasslands, while encouraging fresh shoots, prevents the development of large hummocks under which animals burrow and seek shelter.

Management action

- Extend predator control into arid zone areas wherever possible.
- Monitor spectacled hare wallaby and predator numbers throughout their range.
- Modify burning practices to ensure appropriate shelter is available.

Figure 1.34: Spectacled hare wallaby peering from its daytime nest, often called a 'squat'. (Photo: Jiri Lochman.)

Figure 1.35: Spectacled hare wallaby profile. The orange fur around its eyes certainly look like 'spectacles'. (Photo: Dave Watts.)

Rufous hare wallaby *Lagorchestes hirsutus*

Aboriginals of the Tanami Desert region called these animals *mala*, a name that has been adopted into common usage. The only secure natural rufous hare wallabies populations are on the Western Australian offshore islands of Dorre and Bernier, where in 2010 censuses there were about 1800 individuals on Dorre Island and 1300 on Bernier Island. Here they occupy most habitats such as the dune system, *Triodia* grasslands and heathlands.

The rufous hare wallaby is classified as vulnerable by the 2011 IUCN Red List of Threatened Species. In the Tanami Desert, where they ultimately became extinct in 1991, mala lived in the vast tracts of hummock grasses *Triodia pungens* and *Plectrachne schinzii*, where they were particularly associated with areas of high plant diversity. Even up until the 1920s the unnamed Tanami Desert subspecies of the rufous hare wallaby was distributed widely through the extremely arid zones of central and eastern Western Australia, western Northern Territory and north-western South Australia. However, later there was a steady decline in their numbers, until by the 1980s the

Figure 1.36: Rufous hare wallaby foraging along a sandy dune area on Dorre Island. (Photo: Jiri Lochman.)

subspecies was represented in the wild by a few small colonies in the Tanami Desert. Fortuitously the Parks and Wildlife Commission of the Northern Territory had earlier initiated a captive breeding program. It established a colony in Alice Springs and another near Uluru. By 1992 the last mainland colony in the wild had disappeared because of drought and fire.

Fortunately, in 1998–99 the Western Australian Government's Department of Conservation and Land Management obtained founder mainland mala from captive colonies in the Northern Territory and established a colony on Trimouille Island in the Montebello Island group, and a second colony at Dryandra Nature Reserve near Perth. These bred successfully, and the subspecies has been officially recognised as no longer being 'extinct in the wild' but instead as being 'endangered in the wild'.

The Tanami Desert subspecies is genetically more diverse than the Dorre and Bernier Island subspecies *Lagorchestes hirsutus bernieri*. Consequently they are probably better candidates for future reintroductions. Currently captive populations of rufous hare wallabies have now been established in Francois Peron National Park, Dryandra Nature Reserve, Watarrka National Park, Uluru-Kata Tjuta National Park, the Alice Springs Desert Park and Scotia Sanctuary. The widespread distribution of these self-sustaining populations should ensure that any catastrophic events do not wipe out all populations.

As their common and scientific names suggest, they have a long reddish-brown coat. Bernier and Dorre Island males average 1.6 g and females 1.7 kg. Mainland desert animals weigh less, with males averaging 1.2 kg and females 1.3 kg. This is the reverse of the trend for all other island macropod species, which weigh less than their mainland counterparts. The island rufous hare wallabies have a head-body length averaging 330 mm in males and 375 mm in females. Tail length averages 270 mm in males and 275 mm in females. Pes length is 108 mm and crus length is 135 mm.

They are nocturnal browsers that select plant parts having either high nutritional value (seeds) or high water content (leaves of forbs and roots). They readily eat the occasional arthropod. When necessary they can subsist on extremely low-quality fibrous diets provided by the hummock grasses. They commonly rest up in thickets in a scrape, usually under hummock grass or low heath *Thryptomene* spp. shrubs. Sometimes they rest in a burrow (approx. 1 m long and 0.3 m deep).

Breeding biology data are sketchy. The sex ratio is parity for adults, but is skewed towards males in pouch young 2:1. Sexual maturity is dependent on body condition, ranging from 5–23 months in females and about 14–20 months for males. They are continuous breeders. They are polyoestrous and monovular. Their gestation period is 29–31 days. Embryonic diapause occurs. Between one and two pouch young are born each year. Pouch life is 124 days.

Historic threatening factors to mainland populations

- Predators, notably foxes and feral cats (cats eat both young-at-foot and adults).
- Competing herbivores such as rabbits and sheep.
- Fire regimens altered from the traditional patchwork burning of the Aboriginal peoples to current management practices, where summer burning of large areas became commonplace.

Current threatening factors to island populations

- Genetic inbreeding.
- Probable vegetation changes associated with climate changes.
- Fire.
- Novel environmental challenges, such as disease.

Management action

- Ensure that habitat-threatening factors do not occur on offshore island refugia.
- Continue reintroductions to appropriate habitats and ensure identified threatening factors are controlled.
- Maintain ongoing monitoring of reintroduction populations.
- Ensure that novel diseases are not introduced onto offshore island refugia.
- Ensure appropriate informed management of all areas in which populations occur.
- Maintain high-quality communication between all agencies involved in undertaking and maintaining reintroductions across Australia.

Figure 1.37: Rufous hare wallaby resting in a protected site deep within vegetation on Dorre Island. (Photo: Marie Lochman.)

Wallabies and kangaroos

There are 13 species in this assemblage. They range from the parma wallaby, having a body weight of 3–4 kg, through to the red kangaroo, weighing up to 90 kg. Some species, such as the black wallaroo, have a restricted distribution, while others, such as the euro and the red kangaroo, are extremely widespread. Members of this assemblage are primarily grazers.

Figure 1.38: Eastern grey kangaroos *Macropus giganteus* in the Otway Ranges in Victoria. Note the tightly drawn entrances to their pouches. (Photo: Jiri Lochman.)

Agile wallaby *Macropus agilis*

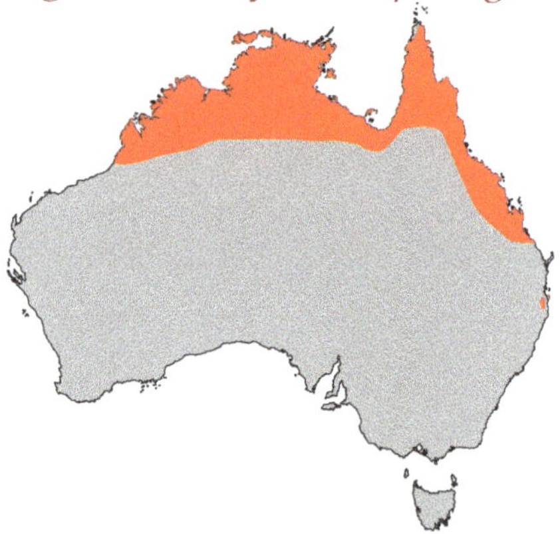

This timid, alert, medium-sized wallaby is common in open woodlands and riverine habitats across the northern regions of Western Australia, the Northern Territory and Queensland. Agile wallabies crossed a land bridge between Australia and New Guinea about 2 mybp. Today the local native people hunt them in the savannah areas and monsoonal woodlands of southern New Guinea. They are still abundant in the Western Province of Papua New Guinea, but are uncommon elsewhere.

Across the more tropical part of their distribution, they primarily graze native grasses during the wet season but switch to include more and more browse over the dry season.

In some areas they severely overgraze their environment, notably riverine habitats. They venture readily onto cultivated land, and in some areas of coastal Queensland this includes cane fields, where their propensity for digging for roots has resulted in farmers shooting and poisoning them. These crepuscular–nocturnal feeders have been observed eating a wide variety of foodstuffs, and in some instances they have adopted quite unusual feeding behaviours.

The following observations from Boodjamulla National Park in remote north-west Queensland illustrate this. Here during the wet season in the riverine habitat they seek out emergent *Livingstonia* palm seedlings (14 cm), grasp the main stem in their incisors and gently pull the whole seedling out of the ground. They then eat the white stem and root but leave the green parts uneaten. Over most of the year they eat the outer fruit of these palms and leave the tough nuts behind. However, in the hard times during the dry season, adults crush the entire seed between their molars. Young animals are unable to do this. In these hard times, agile wallabies eat the seeds of many plants that have passed through the guts of the abundant resident frugivorous birds. They also eat the unpalatable tough pandanus leaves, even dry ones.

Figure 1.39: Female agile wallaby on the woodland floor in Nitmiluk National Park. Note the bulge of a large young 'joey' in her pouch. (Photo: Jiri Lochman.)

Figure 1.40: Agile wallabies in the early evening light on the forest edge of East Point Reserve near Darwin. (Photo: Jiri Lochman.)

A further unusual dietary practice is where young weanling agiles sometimes consume bird droppings. However, agile wallabies will not eat the abundant spinifex grass except immediately after rain when the unpalatable outer resin coat has been washed off.

In some areas of the Top End of Australia, where saltwater crocodiles *Crocodylus porosus* are a constant threat to any medium to large mammal near the water's edge, some agile wallabies appear to have learnt to avoid this risk by digging for water a short distance away from the edge of waterways (billabongs, creeks and rivers).

Agile wallabies have an upper body that is a sandy-brown, and their undersurface is a pale white to cream. They are characterised by a distinct pale facial stripe and a pale rump stripe. The species exhibits sexual dimorphism, with males ranging in weight from 16–27 kg and females 9–15 kg. Head-body length is up to 850 mm in males and 720 mm in females; tail length is up to 840 mm in males and 700 mm in females. They have large ears. The phenomenon of island dwarfism of large vertebrate species can be seen in the population of agile wallabies found on North and South Stradbroke islands that have been separated from the adjacent Queensland mainland for the past 15 000 years. Animals in these populations are about 20 per cent smaller than the nearby mainland animals at the same latitude.

In the wild, sexual maturity is at about 15 months in females and 17 months in males, but is earlier in captive animals. They are continuous breeders, with a postpartum oestrus shortly after giving birth. Their gestation period is 30 days. They have delayed implantation. Pouch life is about 7–8 months, and weaning is 10–12 months.

Threatening factors

Over most of their distribution they are not threatened. In many instances they occur in very high numbers, and in some instances are severely damaging riverine habitats.

Management action

- Control of their numbers in a few areas is important to lessen long-term environmental damage.
- Ensuring appropriate habitat remains in the long term is important.

Antilopine wallaroo *Macropus antilopinus*

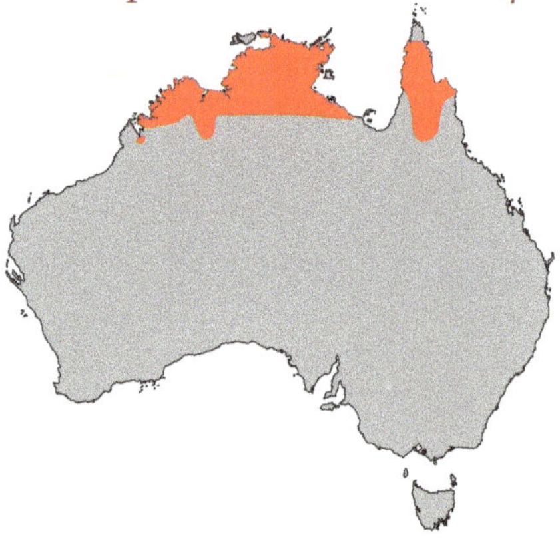

These large macropods, little known to science, are well known to the Aboriginals of the far north of Australia. Western Arnhem Land Aboriginals acknowledge the wallaroo's distinctive sexual dimorphism by calling the males *kandakeet* and females *kandite*.

Antilopine wallaroos are found in open savannah woodlands across the far north of Australia in two disjunct populations, one population in Far North Queensland and the other across the northern regions of the Northern Territory and the Kimberley region of Western Australia. Separating the two populations are the Gulf Plains of Queensland, dominated by semi-arid grasslands and their accompanying red kangaroos.

Antilopine wallaroos are primarily grazers of perennial native grasses on flat and undulating country, but may move into nearby rocky habitats. Normally they are crepuscular, but will emerge from their daytime lie-up during cool dry periods. Even though they are similar in size, they are often sympatric with the closely related euro or common wallaroo. Where this happens, the euros lie up during the day in nearby rocky high ground and come down into the adjacent grasslands to feed.

Figure 1.41: Two antilopine wallaroos illustrating sexual dimorphism. The female is grey above and the male a deep rufous colour.

In recent years antilopine wallaroo numbers have declined, in some areas dramatically, over much of their historic geographic range. This decline is probably a multifactorial event because of changing habitats and, in Queensland, competition with eastern grey kangaroos as they have slowly expanded their range as far as northern Cape York.

Habitat changes caused by land management practices haven't been particularly favourable to antilopine wallaroos. Over large areas infrequent fires have resulted in thick bush-and-tree regeneration, making those areas a less favourable habitat for them. In marked contrast, in other areas burning has been too frequent – sometimes 100 per cent of the land – has been burnt annually. Here plant biodiversity has dropped, and again the quality of the habitat has been lessened. Like the other large kangaroos, antilopine wallaroos are attracted post fire to available green pick, and may aggregate temporarily in large numbers.

Antilopine wallaroos are gregarious and remain in small family groups, each headed by a dominant male. They are sexually dimorphic, with males weighing up to 70-plus kg but more commonly around 55 kg, and females up to 24.5 kg. Head-body length is up to 1200 mm in males and 935 mm in females; tail length is up to 960 mm in males and 815 mm in females. Males usually have a deep reddish-brown pelage while that of females is usually light tan to pale grey. Both sexes have paler underparts.

They are continuous breeders, but most young are born towards the end of the wet season. They do not appear to have a postpartum oestrus. Their oestrus cycle is believed to be 36–44 days and their gestation period is 34 days. They have delayed implantation, but whether they have embryonic diapause is unknown. Pouch life is about 36 weeks.

Males in particular have a characteristic broad 'Roman nose'. This is a result of an enlargement of the nasal chambers and the greater development of the enclosed respiratory turbinates. This may be to minimise respiratory water loss, which is of particular importance for animals that have to live through long dry hot periods when available drinking water is limited.

When ambient temperatures rise towards 40°C, antilopine wallaroos attempt to cool themselves initially by panting through the nose with the mouth closed, but as temperatures continue to rise panting occurs through the mouth. Antilopine wallaroos also cool themselves by evaporation of saliva that they spread over their inner thighs.

Threatening factors

- A little understood decline in their numbers that occurred across much of their historic range in Queensland from the 1980s and 1990s until the present is of concern. Ongoing monitoring is needed to evaluate this.
- Inappropriate fire management.
- Loss of access to permanent water, especially towards the end of the dry season.
- Land alienation.

Management action

- Maintain a monitoring program for the species across its entire range.
- Identify any factors involved in the diminution of antilopine wallaroo numbers.
- Introduce remedial action to address the factors causing declines in population numbers.

Black wallaroo *Macropus bernardus*

These small wallaroos are restricted to the sandstone escarpment of western Arnhem Land where, in some areas, they are relatively abundant. These areas are characterised by open woodlands with a grassy understorey as well as occasional patches of monsoonal forest. Here, the regular traditional mosaic burning of the understorey still practised by the local Aboriginals has kept the habitat similar in form to that created by the Aboriginals using their patchwork burning regimens over the past 40 000–50 000 years.

The Aboriginal people of the East Alligator region (northern boundary of Kakadu National Park) refer to male black wallaroos as *barrk* and female black wallaroos as *djukerre*. They prefer the meat of this species to that of the other kangaroo species in their area. Black wallaroos are still the focus of subsistence hunting by some local Aboriginals.

Black wallaroos are sexually dimorphic, with males weighing 19–22 kg and females up to 13 kg. Their head-body length is 595–725 mm in males and up to 646 mm in females. Their tail length averages 609 mm in males and 575 mm in females. They have a shaggy coat, which in males is dark brown to black, and in females is grey-brown to light grey. They are devoid of facial markings and have a bare muzzle.

Because of their isolated distribution, very little is known about their biology. They are solitary, wary animals. During the day they rest in or under rocky features, leaving these refugia early in the evening to graze on native grasses and forbs. Over the wet season highly nutritious grasses dominate their diet, but as these dry out and become less nutritious they switch to more and more browse. Browse, seeds and fruits supply adequate nutrition over the hot dry season. Indigenous people confirm that black wallaroos eat various species of spinifex, perennial herbs, figs and pandanus found on the rocky escarpments and adjacent regions.

Figure 1.42: Black wallaroo in a shady monsoon forest edge in the dry season. (Photo: Jiri Lochman.)

Little is known of their reproductive biology except that they give birth to most young during the early 'dry' season, but they have the capacity to be continuous breeders.

Threatening factors

- Currently the population estimate for this species is about 10 000 individuals confined to a relatively small area. While the species appears to be stable at present, any significant change to the fire regimen used in its core range could result in a significant decline in the species.
- Overall there is a serious lack of information about their biology, ecology and status.
- Changes to historic fire regimes.

Management action

- This species is in need of proper scientific study.
- Ensure appropriate informed management of all areas in which populations occur.

Black-striped wallaby *Macropus dorsalis*

Commonly called 'scrub wallabies', these cryptic wallabies are common throughout most of their range, from Chillagoe in North Queensland through to northern New South Wales, where small remnant populations still occur on both sides of the Great Dividing Range.

They inhabit forests and woodlands that have a dense shrub understorey. During daylight hours they lie up in groups in semi-permanent camps deep within the understorey. In some areas they lie up in dense thickets of feral lantana (*Lantana camara*). Elsewhere they prefer the understorey to be clear up to a height of 1 m. This allows them good line of sight to see any approaching potential predators.

Historically they were sympatric with the beautiful bridled nailtail wallaby over much of their distribution. However, in modern times they are only naturally sympatric with the bridled nailtail wallabies in protected brigalow scrub habitat near Dingo in central Queensland. Where the two species are sympatric there is no competition for daytime shelter; however, there is some overlap in their diets.

They are highly dimorphic wallabies, where males weigh up to 21 kg and average 14 kg, but females only reach 7.6 kg and average 6.1 kg. Body measures reflect these differences. Males have a head-body length averaging 730 mm compared with females at 520 mm, and tail length in males averages 700 mm and in females 570 mm. Their upper body is a reddish-brown and lower parts are pale white to grey. Their arms and ears are distinctly reddish-brown. They are characterised by a distinct black stripe that runs down the dorsal midline from the neck to the rump and by a distinct pale white hip stripe. They have a pronounced white cheek stripe.

Figure 1.43: Black-striped wallaby proceeding slowly to drink in the evening. These wallabies often move quite a distance from shelter, and therefore are often predated upon by dingoes. (Photo: Jiri Lochman.)

Figure 1.44: Black-striped wallaby drinking. Its distinct black mid-dorsal stripe and white hip stripe are both highly visible. (Photo: Jiri Lochman.)

In the late afternoons, black-striped wallabies begin to move slowly in cohesive groups from their camps towards their night feeding grounds. They have a distinct hopping stance, with their head held low, the back arched dorsally and their arms held forwards and away from the body. Hopping is in short proppy bursts. When the sun sets they move rapidly and directly to the feeding grounds that are often quite a long distance from their daytime shelter. Here they selectively graze grasses (80-plus per cent) and sedges. In the dry season they concentrate on the most favourable patches to lessen energy expenditure from foraging movements. When the nutritional quality of grasses drops, such as during early winter, they shift their foraging effort to browse and forbs. They maximise their feeding time in response to reduced food availability and lower quality.

In areas where they are sympatric with bridled nailtail wallabies, there is dietary competition in spring and winter, when food abundance is low. Under these circumstances the black-striped wallabies are least affected. Dingoes are successful predators of black-striped wallabies, which often graze quite a distance from the safety of adjacent shelterbelts. Even though they historically occurred out towards the fringes of the semi-arid zones in Queensland and New South Wales, they are poorly adapted to living in these areas because they need to drink water regularly.

In the wild, sexual maturity is at about 11 months in females and 16 months in males.

They are continuous breeders. They have postpartum oestrus and their gestation period is 33–36 days. Delayed implantation occurs. Pouch life is 6.4–7.5 months and joeys are weaned between 3 and 6 months after permanent pouch exit. In the wild they live for 10–15 years.

Threatening factors

Although there has been a reduction in the distribution of this species, and in some instances a decline in local populations, overall it is common. In some areas it may be in pest proportions. Dingoes and foxes are significant predators of black-striped wallabies.

Management action

As long as suitable habitat is available the species should not need any special measures to ensure its well-being.

Tammar wallaby *Macropus eugenii*

Francisco Pelsaert, the captain of the Dutch ship *Batavia*, was the first European to see tammar wallabies. Indeed, his report *Ongeluckige Voyagie* in 1629 of 'hopping cats' on the Abrolhos Islands 80 km off Geraldton, Western Australia, is the first written report of any macropod. Today tammar wallabies are one of the most studied of all macropods. Their current status is secure. They have healthy populations on offshore islands ranging from the Houtman Abrolhos to Garden Island and the Recherche Archipelago of Western Australia across to Kangaroo Island in South Australia. Mainland populations in Western Australia, such as those of Tutanning, Boyagin and Perup Nature reserves, are restricted mostly to jarrah (*Eucalyptus marginata*) forests, having dense shrub thickets such as heartleaf *Gastrolobium bilobum* and honey myrtle *Melaleuca viminea*. Usually nearby grassy areas are important for safe grazing. They will eat small amounts of hypogeous fungi. The mainland Western Australian populations are vigorous now that fox control has been in place for several decades. On Kangaroo Island, tammar wallabies have reached plague proportions and in many areas have severely denuded

Figure 1.45: Tammar wallaby feeding in the Perup Nature Reserve in southern Western Australia. (Photo: Jiri Lochman.)

the native vegetation. In such cases some form of tammar wallaby control is essential to allow the vegetation to recover. Feral tammar wallabies on Kawau Island, New Zealand, originating from the now extinct mainland subspecies of South Australia have been used as founder animals to reestablish a mainland population of the South Australian subspecies at Innes National Park on Yorke Peninsula.

At the time of the European occupation of Australia, tammar wallabies had two separate mainland populations: southern South Australia and southern Western Australia. It appears most likely that the South Australian population gave rise to the Western Australian population. Evidence for this comes from evaluations of their tolerance to the plant poison sodium fluoroacetate found in high concentrations in some native plants, notably gastrolobiums. Tammar wallabies from Kangaroo Island have a lower tolerance (lethal dose <0.2 mg/kg) to sodium fluoroacetate than tammar wallabies from mainland Western Australia (>5 mg/kg) and from Western Australian offshore islands (<2 mg/kg). Tammar wallabies probably evolved in South Australia, where gastrolobiums containing sodium fluoroacetate do not occur naturally, hence the tammar wallabies in the area have a low tolerance to the poison. Tammar wallabies that spread to occupy southern Western Australia more than 50 000 years ago, of necessity, evolved a tolerance to the high levels of sodium fluoroacetate found in the extensive stands of gastrolobiums that covered much of the mainland area. Because of the absence of gastrolobiums on the nearby Western Australian offshore islands, tammar wallabies occupying them have not been exposed to the poison since they became isolated from the mainland about 8000 years ago; consequently they have lost much of their inherited tolerance.

Tammar wallabies vary from a greyish-brown through to dark rufous-brown above. Their shoulders and arms as well as thighs are usually rufous. They have grey- to rufous-tinged underparts. They have a distinct white facial stripe that extends to below the eye. A black muzzle enhances this above. Adult male tammars average 7.5 kg and females 5.5 kg. Average head-body length is 643 mm in males and 586 mm in females, and tail length is 411 mm in males and 379 mm in females. Members of island populations in Western Australia are smaller than their Kangaroo Island counterparts.

When hopping at speed they have a bent-over stance, with their head close to the ground. They hold their forelimbs away from the body.

Tammar wallabies are seasonal breeders, with a gestation period of 26–28 days. Most births occur in late January throughout February and into March. Consequently, when the young have their greatest demand for milk (July to October) coincides with when the most nutritious graze is available to the mother.

Female tammar wallabies give birth, have a postpartum oestrus, mate and are usually pregnant within hours. Most adult female tammar wallabies carry a blastocyst in diapause, and have a quiescent corpus luteum throughout the period of lactation. While male tammar wallabies are continuously fertile, other factors, such as peak prostatic development and hormonal status (testosterone in particular), occur from January to March when most adult females enter oestrus. A secondary peak of male sexual preparedness occurs in September and October when the young females of that year come into first oestrous.

Shortly after the summer solstice in December, both the blastocyst and the corpus luteum are reactivated, resulting in the birth of the next young about 19 days later. The joey's first exit from the pouch occurs at about day 190, permanent pouch exit is at about day 250 and cessation of suckling occurs between 300–350 days. Females become sexually mature at 9 months even though they are still suckling their mother at the time. Males become sexually mature at about 24 months. Longevity in the wild is about 11-plus years for males and 14 years for females.

Under conditions of severe environmental water stress, such as during summers when there is limited or no available free water, tammar wallabies have the remarkable ability to survive by drinking seawater. They are able to do so because their kidneys are able to eliminate waste metabolic products and ions using very little water. It appears that of all the small macropods the tammar wallabies have the greatest ability to produce highly concentrated urine. Tammar wallabies also have a remarkable ability to recycle high levels of urea from the blood into the saliva and thus to the stomach. During periods of low nitrogen availability, such as in late summer and early autumn when forage is low in protein, the recycled urea provides a nitrogen substrate for the microbes, which in turn facilitates the efficient fermentation of the available vegetation.

Figure 1.46: A browsing female tammar wallaby with her large pouch young on Kangaroo Island, South Australia. (Photo: Stuart Roper.)

Threatening factors

The main colonies of this species are spread over a large area and numbers appear stable.

- Cats and foxes can be significant predators.
- Ongoing land clearances are a long-term threat.

Management action

- Predator control is essential for the long-term maintenance of successful tammar wallaby colonies.
- Good land management, such as having appropriate fire regimens to ensure the regeneration of shelter thickets, in areas currently holding sustainable populations should ensure viable populations for the future.
- Long-term monitoring of tammar wallaby populations is important.

Western grey kangaroo *Macropus fuliginosus*

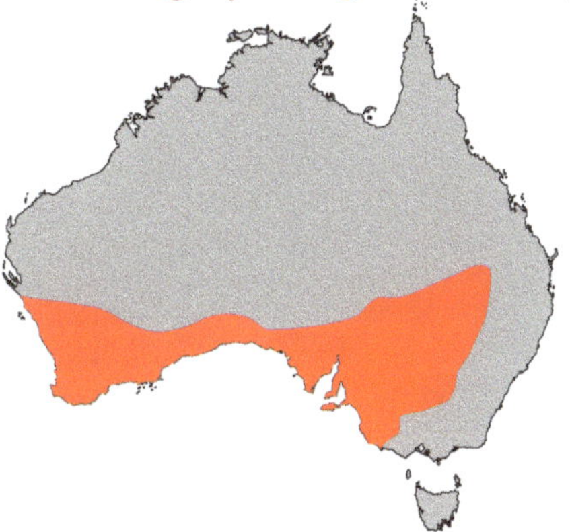

This is an extremely common gregarious species found west of the Great Dividing Range from lower central Queensland to western Victoria across lower South Australia (including Kangaroo Island) and into southern Western Australia, where they may be found as far north as southern Shark Bay. They are found in areas of uniform or winter rainfall. These are mostly southern Mediterranean climatic zones that support xeric plant assemblages. Western grey kangaroos are large kangaroos (females may weigh up to 39 kg and males to 72 kg) that are seen readily at dusk and dawn feeding in open grasslands adjacent to shrubby woodlands where they lie up during the day. In summer, sites where they lie up are frequently characterised by dusty shallow depressions where the kangaroos have dug hip holes for comfort.

They are crepuscular–nocturnal. Under normal seasonal conditions they selectively graze monocotyledons (90 per cent), especially native grasses such as *Arista* spp., *Enneapogon* spp., *Stipa* spp. and secondarily select dicotyledonous plants. In many localities sedges are a major part of their diet. They are well adapted to feeding on and efficiently processing coarse low-quality forage. Even so, they prefer higher quality graze. They are partial to improved pastures, some grain crops and graze among other cropped areas, such as tree farms, orchards and vineyards. They have been observed eating fungi.

They are opportunistic browsers of the widely distributed extremely common shrubby gastrolobium plants in southern Western Australia. Western grey kangaroos of Western Australia have a high tolerance to fluoroacetate, but animals of the eastern mainland distribution do not have this tolerance. The latter group of animals, if challenged with fluoroacetate, either in plants or as the commercially produced poison 1080, are highly susceptible, and usually die by cardiac failure. The difference in susceptibility of the populations suggests that the western grey kangaroo originated in Western Australia and spread east.

Figure 1.47: A mob of western grey kangaroos grazing in the late afternoon in south-western Western Australia. (Photo: Jiri Lochman.)

Figure 1.48: A female western grey kangaroo on the shoreline at Cape Le Grand National Park with a large young at foot suckling. (Photo: Dennis Sarson.)

Over much of their distribution western grey kangaroos are extremely abundant, and are harvested, under a variety of government licensing schemes, for human consumption as well as for pet meat.

Mainland animals are characteristically a brownish-grey above and pale grey underneath. Their face, hands and tail are a dark brown-grey. Their muzzle is covered in fine hair. The tip of the tail is characteristically dark. Kangaroo Island animals are darker than their mainland counterparts. The species is sexually dimorphic, with a head-body length of 950–2225 mm in males and 950–1750 mm in females; tail length is up to 1000 mm in males and 815 mm in females.

Males, in particular, have a distinctive musty odour (curry-like) that is quite apparent even in the shallow scrapes that they make for lying in during the day. This odour is believed to be the result of an exudate secreted from their paracloacal glands. Home ranges vary in size with the quality of habitat. In favourable habitats, such as southern Western Australia, home ranges may be less than 40 ha; however, in the inhospitable arid zone, home ranges may be as large as 700 ha. Of the few animals that disperse, these are mostly young males.

In the wild, sexual maturity is at about 18 months in females and 30–36 months in males, but is earlier in captive animals. As occurs in other large kangaroo species, males may be morphologically and physiologically sexually mature, but are held in check by the dominant males within the group. Females give birth to most young during the summer period (seasonal breeders). Births occur as early as late November, with the peak in January and a few in February and even early March. The timing of most births varies with the level of available good-quality nutrition for the females. With readily available early spring–summer good-quality graze, the births can occur as much as 4 to 6 weeks earlier than is usually the case. Their oestrus cycle is 35 days, and the gestation period is 30.5 days. They do not have embryonic diapause. Pouch life is about 10 months, but they suckle for up to 18 months.

This highly adaptable species is not under threat across its range; if anything, it is expanding its range.

Management action

Sustainable harvesting of this species is important in many areas to minimise long-term damage to the native vegetation in areas where it is found.

Eastern grey kangaroo *Macropus giganteus*

This is one of Australia's most common kangaroos. It is restricted to the eastern side of the country, where it occurs through most of Queensland, New South Wales and Victoria, as well as small populations in southern South Australia and north-east Tasmania (where they are called 'Forester' kangaroos). Mainland numbers are believed to fluctuate between 5 and 10 million animals.

In their western distribution of New South Wales, Victoria and South Australia they are sympatric with western grey kangaroos. Usually they are found either where the summer rainfall is greater than the winter rainfall or where there is no reliable rainfall period. This area is mostly covered by mesic vegetation. However, over the past 30–40 years the increase in the number of watering sites, coupled with the animal's suite of anatomical and physiological adaptations, have allowed the eastern grey kangaroo to have made a major expansion of their range into more arid zones. The main driver of their numbers is the availability of good quality graze.

Even though they are not as well adapted as red kangaroos, eastern greys have kidney structures and functionalities conducive to impressive water-conserving capacities. They also have excellent thermoregulatory capabilities.

Although both species of grey kangaroo hop in a similar manner – with an upright stance and with their tail slightly curved upwards – they can be distinguished from each other by the more robust body form and lighter pelage of the eastern greys.

Over most of their range they live in dry sclerophyllous shrublands and forests, where they shelter during the day. In the more arid rangelands, such as western New South Wales, they are found in mixed chenopod shrubland and grasslands. In most places eastern grey kangaroos spend much of

Figure 1.49: Two eastern grey kangaroos flank a dependent young at foot. (Photo: Dave Watts.)

Figure 1.50: A mob of eastern grey kangaroos resting. (Photo: Dave Watts.)

their time in the taller woodlands with substantial undergrowth, whereas western grey kangaroos seem to prefer lower woodlands and heathlands.

During the day, animals disperse into small family groups to lie up in secluded areas as well as graze understorey vegetation, such as grasses, ferns and shrubs. Towards evening they leave their shelter slowly for their grazing grounds, where many family groups may graze alongside each other. Local environmental and seasonal influences determine the overall group size.

They readily share open grassland feeding areas with western grey kangaroos. Aggregations are often in the dozens, but may occasionally be as many as thousands. Eastern grey kangaroos mostly graze the edges of farmland, but they rarely move more than 150 m into open pastureland. They have overlapping home ranges that vary with environmental conditions, from about 20 ha in optimal habitat to 1000-plus ha in arid western New South Wales. Eastern grey kangaroos appear to be a little less well adapted to feeding on and efficiently processing coarse low-quality forage than western grey kangaroos. Eastern grey kangaroos are attracted to burnt areas, where they graze new shoots. Even in stands of mature grasses eastern grey kangaroos will actively select for green shoots.

This species exhibits sexual dimorphism, with males weighing up to 90 kg (but more commonly 70 kg) compared to about 35 kg for females. Morphological measures are head-body length for males of up to 2300 mm and females 1860 mm; tail length is up to 1100 mm for males and 840 mm for females. Their muzzle has a fine cover of hair. Their upperparts have long, grey to grey-brown fur, and their underparts are a much lighter shade of grey to off-white. Their tail is characterised by a dark tip.

In the wild, sexual maturity varies significantly in populations occupying different habitats and different geographic locations. It is at about 18–24 months in females and 30–36 months in males, but is earlier in captive animals. They give birth to most young during the summer period, but have the capacity to be continuous breeders.

Postpartum oestrus is not present. Their oestrus cycle is 46 days and the gestation period is 35–36 days. Delayed implantation is not present. Embryonic diapause is until the first young emerges from the pouch. Pouch life is about 36–44 weeks, but they may suckle for up to 18 months.

This species has large populations that are able to support sustainable harvesting.

Management action

Sustainable harvesting of this species is important to minimise long-term environmental damage in many areas where they are found in high densities.

Western brush wallaby *Macropus irma*

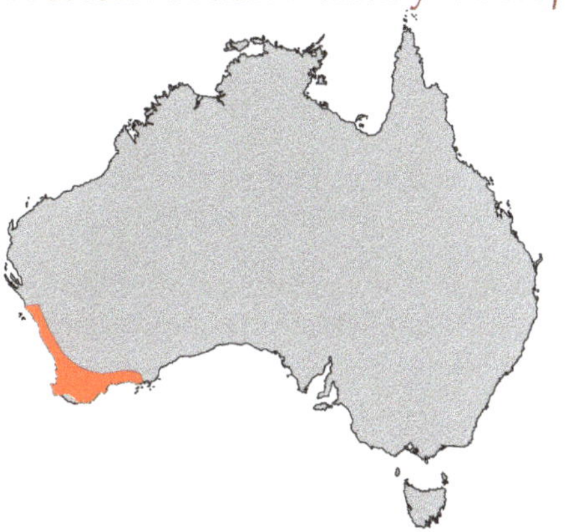

Also called the 'black-gloved wallaby', this pale grey wallaby is characterised by having a white facial stripe that extends from its mouth to the base of its ear, and by ears that have a distinct white inner surface that contrasts with the black of their outer edge. It has black forepaws (gloves) as well as a black tuft on the distal third of the tail.

Historically, Western brush wallabies have been restricted to the southern corner of Western Australia. Shortly after foxes reached there in the early 1900s their predation of juvenile western brush wallabies caused a major decline in western brush wallaby numbers throughout most of the twentieth century. Following several decades of baiting in the late 1900s, fox numbers in Western Australia's southern forests dropped significantly and predation of the newly emerged pouch young decreased.

Figure 1.51: A vigilant western brush wallaby in Wandoo woodland at Dryandra Reserve near Perth, Western Australia. (Photo: Jiri Lochman.)

Consequently the number of western brush wallabies rose dramatically. Even so, they are restricted to open woodlands (dry sclerophyllous shrublands and forests) that have a low clumped understorey. Although they commonly frequent gullies and stream edges, they are absent from the much wetter areas (annual rainfall less than 1000 mm) of south-western Western Australia. In the late afternoon they proceed to their preferred foraging areas, hopping quickly with their head held down and tail extended directly behind. They are predominantly (79–98 per cent) browsers of the leaves of low shrubs and to a lesser extent forbs and herbs. They appear to eat readily gastrolobiums, acacias, cycads and hakeas.

Sexual dimorphism is absent. They weigh 7–9 kg, head-body length is about 900 mm and tail length is 600–950 mm. Western brush wallabies are highly territorial. If relocated out of their familiar territory they panic. Hence attempts to study them in captivity have not been successful. Therefore little is known of their reproductive biology.

They appear to be seasonal breeders because most young are born in winter, but they are believed to have the capacity to breed at alternate times if the prevailing environmental conditions are favourable. Western brush wallabies wean their pouch young at a very early age, usually when the young is about 1 kg. It appears that they become independent as soon as they exit the pouch. Recently weaned females associate with their mothers, giving them a little protection. However, the young males disperse immediately after pouch exit and consequently are quite likely to die from predation.

Threatening factors

- Fox predation can be extremely damaging to a population.
- Loss of suitable habitat, particularly by land clearances, is a long-term threat.

Management action

- Maintain long-term fox control.
- Maintenance of appropriate habitat should ensure viable populations.
- Instigate monitoring of their population across the range of their habitats.
- Further field investigation of their biology is important.

Parma wallaby *Macropus parma*

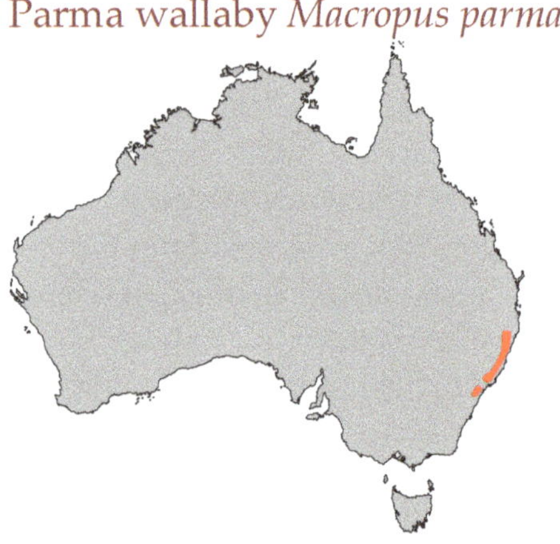

These are small wallabies having a discontinuous distribution along the Great Dividing Range and adjacent coastal areas of northern and central New South Wales, where they live primarily in the wet sclerophyllous forests.

They are nocturnal. During the day they shelter in dense thickets, and emerge in the late afternoon or early evening to graze on the edges of open patches of grassland. Parma wallabies are mycophagous, but what contribution fungi make to their overall diet is not known.

When hopping at speed they have a bent-over stance, with their head close to the ground and their forelimbs close to the body. When they hop, their tail is held high in an upward curve. Their pelage is greyish-brown to reddish-brown above. They are paler underneath, with a white throat and chest. They have a distinct white facial stripe that extends below their eye.

They exhibit sexual dimorphism, with males ranging in weight from 4.1 to 5.9 kg and females 3.2 to 4.8 kg. Head-body length is 450–530 mm and tail length is 405–545 mm.

Figure 1.52: A parma wallaby standing observing from the forest verge. (Photo: Jiri Lochman.)

In Australia, wild male parma wallabies become sexually mature at about 20–24 months and females from 12 months. They are continuous breeders, with a tendency for most births to occur between February and June. They do not have a postpartum oestrus. They have a gestation period of 35 days, and delayed implantation is usual. Pouch life is for about 7–8 months, and weaning is completed between 10 and 11 months.

Threatening factors

- Predation by feral cats and foxes.
- Loss and fragmentation of habitat through land clearing.
- Removal of understorey and shrub layers by domestic stock.
- Too frequent burning of the understorey and shrub layer, especially at forest margins.
- Misadventure; vehicle accidents.

Management action

- Predator control is important for the long-term well-being of parma wallaby colonies.
- Retention and revegetation of forest edges is important.
- Undertake the appropriate use of fire.
- Developing larger suitable habitats and movement corridors between nearby suitable habitats is critical.
- Monitor the populations and take appropriate action as needed to enhance the sustainability of these populations.

Figure 1.53: A parma wallaby foraging in a forest opening. (Photo: Jiri Lochman.)

Whiptail wallaby *Macropus parryi*

Commonly called the 'pretty-face wallaby', as the name suggests, this is an extremely attractive animal that has a relatively long tail (greater than the head-to-body length). Whiptail wallabies are gracile animals, with characteristic white muzzle and hip-to-rump stripes as well as submarginal white ear stripes. Their body coat colour is pale, being greyish in winter and browner in summer. Their elongate tail is pale in colour, except for a dark tip.

This species occurs in open forest with a grassy understorey from far north-eastern Queensland down to northern New South Wales. They are crepuscular–nocturnal. Groups are seen often in the early morning or late afternoon on hilly slopes either grazing on native grasses and perennials or hopping away rapidly. They rarely drink.

Whiptails are a gregarious species that commonly form social groups of up to 30 individuals and, rarely, up to 50 individuals. Groups appear to be controlled by a single dominant male.

They exhibit sexual dimorphism, with males ranging in weight from 14 to 26 kg and females 7 to 15 kg. Head-body length is up to 1005 mm in males and 880 mm in females; tail length is up to 1050 mm in males and 860 mm in females. In the wild, sexual maturity is 18–24 months in females and 24–36 months in males. They are continuous breeders. They do not have a postpartum oestrus. Oestrus is 38–44 days, and the gestation period is 36–41 days. Delayed implantation is usual. Pouch life is about 9 months, but suckling can continue until 15 months.

In southern Queensland, where they are abundant, there may be up to a million individuals over their current range. In the 1980s they were harvested commercially.

Management action

As long as appropriate grasslands on hilly well-forested country are available, this species should not be under any serious threat, so no particular management is needed for the survival of this species.

Figure 1.54: A large male whiptail wallaby in open forest. Note his long powerful tail. (Photo: Jiri Lochman.)

Figure 1.55: A female whiptail wallaby with her dependent young at foot. Their attractive facial features are responsible for their alternate name 'pretty-face wallaby'. (Photo: Hans and Judy Beste.)

Euro or common wallaroo
Macropus robustus

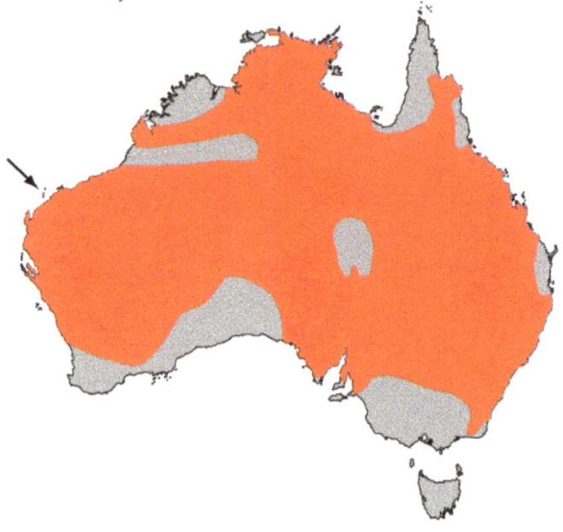

This, the most widespread of all the macropods, has three subspecies: *M. robustus robustus* of the eastern non-arid zone; *M. robustus erubescens* of most of the rest of Australia to the west of the Great Dividing Range; and *M. robustus woodwardi* of the Kimberley and north-western Northern Territory. An apparent fourth subspecies *M. robustus isabellinus*, restricted to Barrow Island (arrowed) was recommended for synonomy with *M. r. erubescens* in 2004.

In the eastern parts of its range, they are called the 'common' or 'eastern wallaroo' but in the western part of their distribution they are known as 'euros'. They are heavy-set grazers that thrive in a vast array of habitats, such as the desert–arid zone of central and western Australia, the far northern area of Australia, under the influences of monsoonal wet seasons, through to the alpine areas of the Victorian–New South Wales border region. These animals prefer rocky habitat (steep escarpments, rocky hills, stony aggregations) within their home ranges, where they can lie up during the day.

In extremely hot periods they often crawl deep within piles of boulders to avoid the ambient heat of nearby open areas. Euros are absent from the extremely inhospitable desert areas south of Broome in Western Australia and central Australia. They are also absent from the escarpment areas of the Kimberley where several other small macropodids live. In Arnhem Land they are frequently found living in areas adjacent where their larger cousin the black wallaroo lives. Here the black wallaroo is generally found in the higher escarpment country while the euro lives lower down near the flatlands.

Euros are a relatively solitary, sedentary species that are nocturnal-crepuscular grazers of native grasses and forbs. Over much of their range their sedentary lifestyle has necessitated their being able

Figure 1.56: A common wallaroo on a rocky slope in the Flinders Ranges of South Australia. (Photo: Jiri Lochman.)

Figure 1.57: Euro among spinifex grass on Barrow Island. These animals are significantly smaller than their mainland counterparts. (Photo: Jiri Lochman.)

to survive on less nutritious plant material than their sympatric relative, the red kangaroo. They can exist solely on a diet of spinifex. However, in the tropics they graze primarily on nutritious grasses over the wet season, but become more reliant on browse over the dry season.

The pelage of this species is highly variable. The pelage of the eastern form is long and dark grey in males and a much lighter shade of grey in females. Animals in the far north-east have a shorter coat that is a dark rufous-rusty colour in males and a lighter shade of rufous through to a blue-grey in females. Animals of the Top End are a pale fawn colour. Euros, particularly those of the north-western region of Western Australia, have a distinct dark rusty-red pelage. Although their pelage colour may vary greatly, all euros–common wallaroos have a characteristic large hairless rhinarium and dark paws and feet. Their tail is not normally tipped black, as is the case in the grey kangaroos. They have an upright bounding gait.

Sexual dimorphism is readily evident. On the mainland, males weigh up to 60 kg and females up to 28 kg. Head-body length is up to 1990 mm in males and 1580 mm in females; tail length is to 900 mm in males and 750 mm in females. For decades the small size of the euros on Barrow Island (population approx. 1800), where males only average 18 kg and females 8.6 kg, gave rise to them being considered as a distinct subspecies. However, they are not; instead, they are a good example of island dwarfism.

About 8000 years ago Barrow Island was separated from the nearby mainland by rising sea levels. The island was, and still is, dominated by hummock grasslands that, while abundant, are of low nutritional quality. Under conditions of low or declining nutritional intake an animal's body condition will decline. The principal avenue for an animal to redress these circumstances is to seek out higher quality forage. If this is not possible, the only solution available to the animals is to increase foraging time in an endeavour to obtain more nutrients, albeit in an ineffectual way. Being a relatively large species, euros of the Barrow Island founder population were nutritionally stressed, simply because they could not get enough quality food to maintain their body condition. Over time, genetic inbreeding occurred, which in turn aggravated the population's abilities to respond to reaching the limit of its food supply. Ongoing nutritional stress stimulated selection for a smaller body form in this euro population.

The arid-adapted mainland form of euro (*M. robustus erubescens*) has the lowest maintenance requirements for nutrients of all kangaroos studied to date. In periods of drought it can survive on the 'soft' spinifex *Triodia pungens*, which has a nitrogen content (dry weight) ranging from 0.06 to 0.5 per cent.

Figure 1.58: Euros drinking at a windmill overflow in Cape Range National Park. Even in extremely arid areas euros appear to need to drink several times each week. (Photo: Jiri Lochman.)

Figure 1.59: Euro digging in a creek bed for water on Barrow Island. (Photo: Jiri Lochman.)

The eastern form, *M. robustus robustus*, usually called the 'common wallaroo', is less well adapted, and its maintenance requirements are similar to those of other comparably sized macropods, such as the eastern grey kangaroo.

The euro also has a much lower need for digestible nitrogen (approximately 33 per cent less) than the common wallaroo. Euros, in particular, have a great ability to recycle urea through their saliva to the forestomach, thus lessening their need for dietary nitrogen, as well as lowering the amount of water needed to flush it from their kidneys. Arid-zone adapted euros need about one-third less water than the mesically adapted common wallaroos of the eastern forests. The large colon of the euro is 65 per cent longer than that of the common wallaroo, hence the euro has the capacity to absorb more water. Consequently euros produce faeces that are about 40 per cent dry matter, compared with 30 per cent in the common wallaroos. Professor Ian Hume, who has undertaken extensive nutritional studies of kangaroos, suggests that the euro may preferentially retain water within the alimentary tract, thus providing a better environment for microbial digestion.

If euros can obtain water from standing vegetation they do not need to drink. Barrow Island euros eat spinifex that has a low water content, so consequently they have to drink daily from standing water. Euros have a lesser ability to concentrate urine than do red kangaroos. This is possibly because the daily heat loads on a red kangaroo are higher in the summer, and because they seek shade rather than lying up in rocky shelters during the heat of the day.

In the wild, sexual maturity occurs at about 18 months in females and 24 months in males. Most young are born during the summer. However, the females have the capacity to be continuous breeders. Oestrous varies from 34 days in common wallaroos to 45 days in euros. They have a postpartum oestrus. The gestation period of the common wallaroo is 31–34 days and for the euro is 32–35 days. They have embryonic diapause until the young emerges from the pouch. Pouch life is from 37–38 weeks, but young-at-foot continue to suckle up to 64 weeks of age.

Euros of Lake Argyle swim from island to island. Interestingly, when swimming they do so using their legs independently.

Management action

This species is widespread in Australia and occupies many differing geographic zones. It is a common species, and in some instances becomes superabundant, densities of up to 80 animals per square kilometre having been recorded. No particular endeavours to maintain their status across their range appear necessary.

Red-necked wallaby *Macropus rufogriseus*

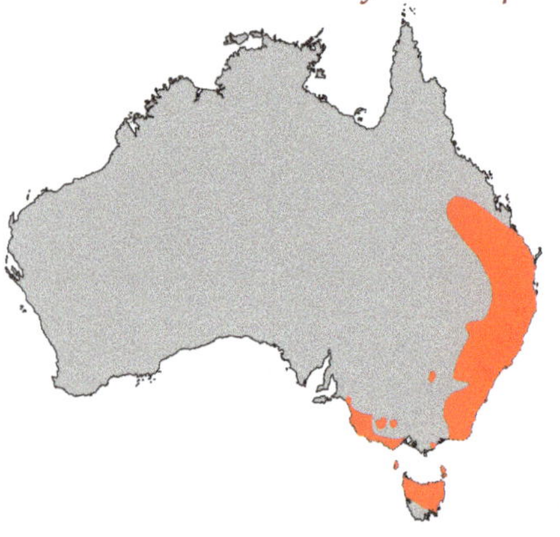

This species is often called 'Bennett's wallaby', especially in Tasmania. They are large forest wallabies that occupy scrub country in a strip from Rockhampton in central Queensland down to north-eastern Victoria. There are a few isolated coastal populations across Victoria and into south-eastern South Australia, where they have a wide distribution but are few in number. A second group of populations occurs in Tasmania and the larger Bass Strait islands. These animals are morphologically and physiologically distinct from the mainland populations.

Male red-necked wallabies may weigh as much as 27 kg and females 16 kg. Head-body length measures up to 890 mm in males and 840 mm in females. Tail length measures up to 875 mm in males and 790 mm in females. Their dorsal coat colour is brownish-grey to a red-brown with a distinct reddish to orange-brown neck. A distinctive white cheek stripe on the upper lip, whitish underparts and black paws and feet characterise red-necked wallabies. The tail is pale. Mainland animals have a short coat, but island animals have a longer, shaggier coat.

Figure 1.60: A female red-necked wallaby with a large pouch young, both grazing in the Bunya Mountains of Queensland. (Photo: Jiri Lochman.)

Red-necked wallabies are found in a variety of habitats, but they live primarily in sclerophyllous forests that have a dense understorey. Larger populations occur where forested areas adjoin cleared areas. The wallabies use the forested areas for secure daytime rest sites and the open areas in the late afternoon and evening for grazing. They are selective grazers of native grasses, and utilise forbs as a secondary dietary component. However, in many areas across their range they are regarded as pests because they heavily graze crops and improved pasture. Harvesting of red-necked wallabies

to supply the fur trade occurred until recent times. In 2000 the commercial harvest quota from Flinders Island was set at 7000. However, this harvest ceased in December 2002.

In the wild, sexual maturity is at between 14–17 months in females and 21–24 months in males. Mainland animals are continuous breeders, but island animals have a restricted breeding season, with peak births between February and April.

They have a postpartum oestrus, an oestrus period of about 33 days and a gestation period of 30–31 days. Embryonic diapause of island wallabies is controlled by photoperiod, in this case increasing day length following the autumn equinox. Under favourable circumstances the blastocyst can remain viable for up to 11 months before it implants. Pouch life is 36–40 weeks, but suckling may continue for a further 10 to 13 weeks following permanent pouch exit.

Management action

Retention of appropriate habitat in Government control.

Figure 1.61: Red-necked wallaby in the alpine region of Tasmania. These wallabies are well adapted to surviving these damp and often snow-bound areas. (Photo: Jiri Lochman.)

Red kangaroo *Macropus rufus*

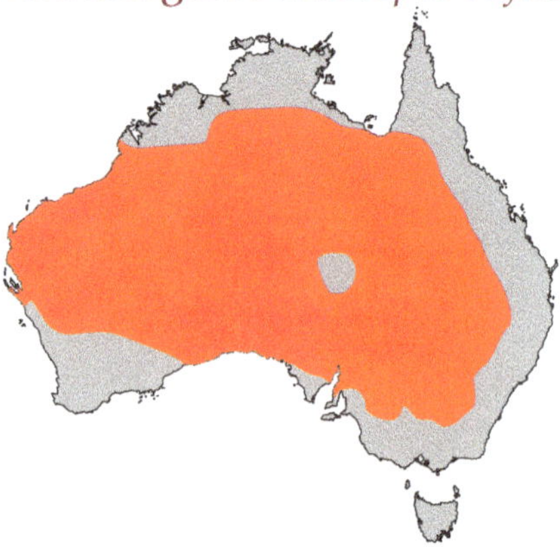

This gregarious species is possibly the most recently evolved of all the kangaroos. Its fossils are found in the Pleistocene deposits, and it appears to have evolved in parallel with the gradual increasing aridity and decline of wet forests characteristic of Australia since the Pleistocene period.

Nowadays it is a common and widespread inhabitant of the arid zone of less than 250 mm rainfall per annum and semi-arid zones with an annual rainfall of 250–500 mm (grasslands, mallee shrublands, mulga scrub, salt plains and deserts), and as such is extremely well adapted morphologically, physiologically and behaviourally to survive in these harsh and demanding environments.

The widespread provision of stock watering bores may have allowed the species to increase its distribution; however, there are marked differences in how much water individuals need. Some individuals appear to drink little and often if water is readily available, and nearby, particularly in the hottest times of the year. Other individuals may drink rarely or at intervals of 4, 5 or 6 days, and travel several kilometres or even up to 10 km to visit open water. Such visitations are more frequent in late summer, especially if no rain has fallen for some time. As long-distance travel is energetically demanding, especially during hot periods, an individual kangaroo can be placed in a dilemma of whether or not to drink. Even though red kangaroos are capable of long-distance travel they are actually quite sedentary. Although their home ranges will enlarge in times of drought, it appears that when favourable conditions return an individual's home range shrinks back to its previous dimensions.

They are large, sexually dimorphic kangaroos where males may grow to 85–90 kg and females

Figure 1.62: Red kangaroos drinking in Sturt National Park. The smaller blue–grey female is colloquially known as a 'blue flyer'. (Photo: Dave Watts.)

to 35 kg. Head-body length is up to 1400 mm for males and 1100 mm for females; tail length is up to 1000 mm in males and 900 mm in females. They are characterised by having a head often described as resembling that of a mule deer. They have large pricked ears, a hairless nasal area and a diagonal white stripe from the corner of the mouth to just below the eye. Their pelage colour varies greatly across their entire geographic distribution, with animals from the eastern edge of their range having a strong colour dimorphism, with males being rusty-brown and females a blue-grey above, each grading to lighter shades below. However, true desert dwellers from the western edge of their distribution usually have both sexes being a deep rusty-brown colour. Their underparts and the distal parts of their limbs and tail are pale cream to grey.

Animal abundance is related to the biomass of available preferred pasture species. Animal numbers rise and fall according to variations in rainfall, which dictate pasture growth. Red kangaroos prefer high-quality graze of short grass, especially green shoots following fires or recent rains, but as the green shoots become scarce they include more and more forbs in their daily diet. Cattle grazing in the arid and semi-arid zones actually creates a subclimax pasture by the stimulation of the grasses to form fresh green shoots that are the main item in the diet of red kangaroos. This has allowed red kangaroo numbers to rise and the species to move into areas where previously it was absent.

Red kangaroos will travel long distances (30 km) to access nutritious succulent grasses. This ability has led to their being locally nomadic. It is only when all green feed is gone that red kangaroos will eat the less palatable dry grass. Given a choice, they avoid areas covered in tall grasses. In times of extreme drought or overgrazing, forbs and chenopods become dominant in their diet. It appears that large males feed on less nutritious plant material than the much smaller females. Natural deaths occur with old age, drought and extreme heat. Fecundity and early juvenile survival are lowered by decreasing food availability as well as by increasing kangaroo numbers.

In the wild, sexual maturity is at about 24 months in females and up to 36 months in males. Male red kangaroos secrete a water-soluble red powdery substance, believed to be cinnabarinic acid, from their skin glands onto their chest and neck during the breeding season. It is thought to be linked with reproductive and social status. To a lesser extent the secretion may be produced by the females. They give birth to most young during the spring–summer period, but they have the capacity to be continuous breeders. Oestrus is 35 days and the gestation period is 33 days. They have a postpartum oestrus, and usually mate within 1 to 2 days of giving birth.

Embryonic diapause continues until the previous pouch young leaves the pouch or is lost through misadventure. Pouch life is about 8–9 months, but the young-at-foot may continue to suckle for an additional 3 months. When environmental conditions are severe, females may become anoestrous. Alternately, they may give birth to a succession of young that die shortly after birth. In the second scenario, the female always has a quiescent embryo available for any improvement in the environmental conditions that allow a pouch young to be raised to leave the pouch. The second scenario is successful because it is energetically inexpensive to produce and replace very small pouch young.

In environments where dingoes are relatively common, red kangaroo numbers can be limited by predation, mostly of young-at-foot. Such predation is particularly evident when available alternate small prey has declined dramatically. Under those circumstances dingoes change their hunting behaviour from solitary endeavours to catch small game such as rabbits to pack behaviour so that they are able to tackle larger animal such as red kangaroos. During drought, the weakening of individual kangaroos as well as their increasing density closer to water points also increases the possibility of their being predated by dingoes.

Professor Terry Dawson of the University of New South Wales spent much of his long illustrious research career studying large kangaroos, and red kangaroos in particular. He reports that the most physiologically comfortable hopping speed for red kangaroos is 15–35 km h^{-1}. Over this speed range they can alter their speed without increasing their energy use by maintaining the same hopping rate but varying the stride length. To achieve speeds of above 35 km h^{-1} they must increase stride length as well as hopping frequency. They can reach a maximum speed of about 70 km h^{-1} and maintain this for a short distance; however, this requires the expenditure of a great deal of energy.

Figure 1.63: A large male red kangaroo hopping slowly though typical hot dry habitat of central Australia. (Photo: Dave Watts.)

Their quite obvious ability to move long distances has resulted in the widely held belief that they will move exceedingly long distances over very short periods to travel to find green pick following recent rains, especially following droughts. It is more likely that the sudden appearance of large numbers of kangaroos on green pick following local storms or fire is the result of local movement (up to 30 km) of kangaroos from the surrounding area.

However, individual red kangaroos have been recorded to travel long distances. According to banding records, a 27+-year-old large male was shot near the NSW–SA border, having travelled 300 km in a direct line from where it had been originally banded. Another in the Wiluna area of Western Australia was reported to have travelled 338 km in 15 weeks. These movements were not responses to fresh green pasture.

In the true arid zone, home ranges may be as large as 800 ha, but in more hospitable habitats they may be as little as 18 ha. Red kangaroos are sympatric with euros over much of their range, and both species use a suite of physiological, morphological and behavioural adaptations for living in the arid zone. However, each species is adapted in slightly different ways. Red kangaroos have a better capacity to concentrate urine than do euros. During periods of extreme heat, reds seek shade under a tree or bush, whereas euros move into shady crevices or caves or under rocky overhangs.

In extreme conditions of hot weather (40°C), red kangaroos in particular lick their forearms from where the saliva evaporates, cooling the blood in the extensive network of blood vessels lying immediately beneath the skin. This cooling technique is effective as long as the animal has access to drinking water. If water is limited, the animal uses this technique sparingly to minimise water loss. Panting is another successful method of cooling the body. The fur of a red kangaroo reflects up to 30 per cent of ambient solar radiation, giving it more protection from solar radiation than does the fur of euros. A red kangaroo's fur provides better protection against heat loss in winter than that of a euro.

This species has expanded its range greatly and in many places becomes seasonally superabundant.

Management action

Sustainable harvesting is necessary to prevent long-term degradation of native vegetation by it in many areas.

Bridled nailtail wallaby *Onychogalea fraenata*

This is one of Australia's rarest macropods. In the early days of European settlement, it was relatively common in the acacia scrubland and grassy woodlands of the semi-arid zone along the western slopes of the Great Dividing Range from northern Victoria to Charters Towers in northern Queensland.

However, dramatic changes to habitats lowered numbers so far that by 1930 the species was presumed extinct. Fortunately, in 1973 a small population was found in Queensland near Dingo, about 150 km west of Rockhampton on Red Hill and Taunton cattle stations. Subsequently, between 1979 and 1984 the Queensland Government purchased both properties, about 11 470 ha, and these were gazetted as Taunton National Park (Scientific). In 1991 the first Recovery Plan for the bridled nailtail wallaby was gazetted, and has been followed subsequently by ongoing Recovery Plans.

Taunton National Park has open grassy eucalypt woodlands (*Eucalyptus populnea*) and brigalow *Acacia harpophylla*. Transition zones between these two plant

Figure 1.64: Bridled nailtail wallabies sniffing encounter. (Photo: Jiri Lochman.)

assemblages are the preferred sites of the bridled nailtail wallabies.

This is a small macropod, with adult males weighing up to 8.3 kg and females up to 6 kg. The head-body length is 430–700 mm. The tail is 360–540 mm. The overall body colour is a grey-yellow with a rufous tinge to the neck, shoulders and upper back. There is a dark mid-dorsal stripe down the back. A white broad crescent stripe runs from behind the ears along the shoulder to behind the forearm. Underparts are cream. These wallabies have a distinct pale facial stripe. Their tail is tipped by a black 'nail', which is a horny process about 3–6 mm long. A terminal tuft of black hair hides this.

The home range of males is about 60 ha and for females about 25 ha. These home ranges have a daytime resting core area and a nocturnal feeding area. During the day these nocturnal wallabies lie up alone in shady scrapes in areas of thick vegetation or in hollow logs. The Queensland Environmental Protection Agency uses this behaviour in its management of colonies under its care by providing large concrete pipes within their territories for them to use.

Towards evening the wallabies browse in the immediate area of their daytime refugia, then slowly make their way to their feeding grounds, which may be a long distance away. They positively select for chenopods, pigweeds and daisies. They browse opportunistically seed heads and leaves (high nutritional value) and normally avoid grasses (high fibre content). However, when there is a shortage of food they will switch to grazing the less nutritious grasses. Under extremely poor conditions they have been observed to supplement grasses by eating fallen leaves of higher nutritional quality, such as the false sandalwood *Eremophila mitchelli* and native myrtle *Myoporum montanum*. Unlike many species of kangaroo, bridled nailtail wallabies do not warn conspecifics of potential danger by stamping their feet.

In this sexually dimorphic species, access to females is greatly influenced by male body weight. In the wild, sexual maturity varies greatly, from 18–35 weeks in females to 34–60 weeks in males. They are continuous breeders and have an oestrus cycle of 36 days. Postpartum oestrus is absent. Their gestation period is about 23.5 days. Delayed implantation occurs. Pouch life is between 16 and 18 weeks. Weaning is about 10 to 15 weeks later.

Conservation status

Endangered. According to the 2011 IUCN Red List of Threatened Species, they are endangered because their geographic distribution is less than 5000 sq. km, and there is a continuing decline of their habitat.

Threatening factors

- Predation by dingoes, foxes and feral cats.
- Competition with feral and domestic herbivores.
- Clearing and burning of remnant habitat.
- Prolonged drought.
- Wildfire.
- Disease and parasites.
- Exotic weed invasion.

Management action

Successful translocations of bridled nailtail wallabies to Idalia National Park, Avocet Nature Reserve, Scotia Sanctuary and Western Plains Zoo have formed the basis of additional breeding populations. This lessens the possibility of extinction, because there are now several geographically separate breeding colonies. If any one colony were extinguished by catastrophe, there should still be viable alternate colonies from which one could realistically reintroduce the species back to the wiped-out site.

This species is the subject of a dedicated Recovery Program by the Queensland Government. The strategies that have been put in place to address the threatening factors delineated above have been extremely successful, and should ensure the ongoing survival of the bridled nailtail wallaby.

The spread of the introduced pasture, buffel grass *Cenchrus ciliaris*, through much of the area in which these wallabies occur naturally is an evolving problem. In summer this grass forms extremely dense dry stands that are prone to very hot fires. The nailtail's preferred habitat, brigalow, is quite sensitive to fire, so excluding fire using wide buffer zones helps maintain the brigalow refugia. Managers can use fire as a conservation tool shortly after rain and during the growing period, when the fires will be cool to reduce fuel loads. Grazing the buffel down first is a possibility; however, one must weigh up the impacts of cattle versus fire. Fire also destroys the hollow logs in which many bridled nailtail wallabies shelter.

Figure 1.65: Bridled nailtail wallaby. These wallabies rarely move far away from adjacent shelterbelts. (Photo: Jiri Lochman.)

Northern nailtail wallaby
Onychogalea unguifera

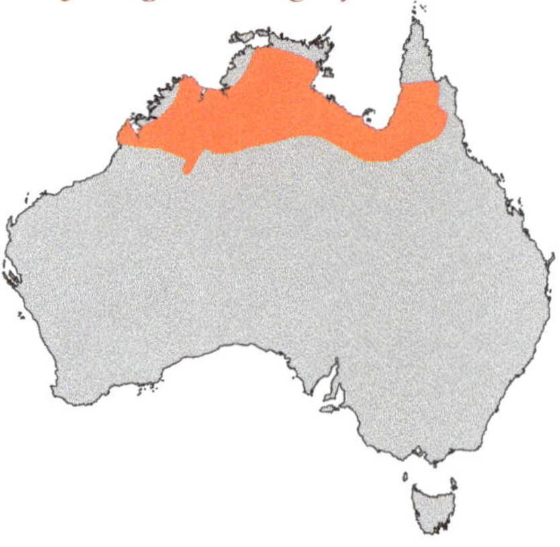

These small wallabies have a wide but discontinuous distribution through the tropics and subtropics of northern Australia. Populations lie between the coastal zone and the 500 mm isohyet. They occur in a variety of woodland types, with associated understoreys of grass through to open coastal plains. They are found along most of the riverine flood plains that crisscross their area of distribution. In most habitats they rely on dense thickets to lie up in during the day. The best populations of northern nailtails occur in heavily grazed granite-based pastoral lands because these areas have abundant herb cover that is the preferred diet of the species. Even though they may be common, they are difficult to observe.

Figure 1.66: Northern nailtail wallaby near pasture. (Photo: Jiri Lochman.)

Figure 1.67: Tail tip of a northern nailtail wallaby. Why these wallabies have this inert extremity to their tail is unknown. (Photo: Jiri Lochman.)

They have a light sandy-brown pelage with a pale cream hip stripe and facial stripe. They have a dark stripe running from the dorsal neck to the tail base. The basal half of the tail is a light grey that changes to a grey-black over the distal half. Usually the distal half of the tail is characterised by alternating light and dark rings. As would be expected from its common name, there is a short (20 mm long by 10 mm wide) horny spur on the tail tip hidden by long terminal black hairs.

They weigh 4–9 kg, their head-body length is 490–700 mm, and the tail length varies from 600–730 mm. This is a species where the tail is considerably longer than the head-body length. They hop with their head held low, their tail turned up in a U-shaped arc and with their arms stretched down and out. As they hop they move their arms in a circular fashion, an action that gave rise to an early common name of 'organ grinder'.

When approached, for instance, by a rider on horseback, they have a habit of freezing, relying on their cryptic coat colour to remain unobserved. They remain stationary until the horse is virtually about to step on them, then they burst out from cover at the last moment emitting a loud 'wut wut' call. Obviously this behaviour startles the horse, often causing the rider to be thrown off the mount. Therefore they are not popular with many station hands.

These nocturnal wallabies are usually solitary, but small aggregations of up to six animals may form at feeding sites. They are primarily browsers that prefer herbs, but may eat succulents and fruits, and under extreme conditions they will eat green grass shoots.

Figure 1.68: Northern nailtail wallaby on a boulder. Note the striped appearance of the terminal half of its tail. (Photo: Jiri Lochman.)

They are unusual among the macropods in having long slender incisors, with the corner incisors being smallest, then the lateral incisors followed by central incisors.

In the wild, sexual maturity is at about 11–12 months in females and 13–15 months in males. They are continuous breeders. They have a postpartum oestrus. Their oestrus cycle is 20–23 days and their gestation period is 17–24 days. They have an embryonic diapause and delayed implantation. Pouch life is about 5–6 months.

Threatening factors

- Changing management practices of pastoral lands, especially the poor use of fire, is potentially a problem.
- If foxes become more common north of their current distribution they could have serious detrimental effects on the population.

Management action

- As little is known of this species, research into its biology and ecology is needed.
- In particular, more details of its habitat requirements are essential.
- Improved management of pastoral lands, including less reliance on late season fires.
- Censuses to determine its status across its geographic range are essential to establish whether or not specific measures for its conservation are needed.

Rock wallabies

It is probable that rock wallabies evolved from thylogale-like ancestors when these were able to successfully utilise the rocky outcrops found under the cover of closed forests. Taxonomically the rock wallabies are also closely allied to the tree kangaroos, and it is probable that both genera split from a common ancestor between 7 and 5 mybp.

By about 4 mybp the ancestral rock wallaby was well established, probably in far north Queensland. At some time between 3.7 and 0.5 mybp there was a rapid radiation of rock wallabies, resulting in the current extant species being found in suitable habitats over most of Australia. It is most likely that in those times rock wallabies occupied much more diverse environments than they do currently.

Over the period 3.7–0.5 mybp there was a gradual drying of the Australian continent, which caused significant changes in the flora of the country, namely a withdrawal of rainforest from much of Australia coupled with an increase in area covered with arid-adapted plant communities. This, with a variety of other factors such as interspecific competition, forced the rock wallabies to withdraw from their earlier diverse environments and retreat to rocky habitats.

In the first phase of their radiation the original ancestors (probably a *Petrogale persephone*-type) are believed to have spread from the closed forests of north-eastern Queensland across similar habitat throughout mainland Australia. Towards the end of this radiation the ancestral *lateralis* group evolved in the western regions of Australia. It is believed that this group became widespread over the western half of the continent, and that one of its offshoots to the Top End formed the basis of the *brachyotis* group that over time evolved into three species. A second offshoot recolonised the east, probably via a southern route, and possibly simultaneously via a northern route to form the basis of the *penicillata* complex.

Morphologically most rock wallaby species are similar to each other, some being indistinguishable in the field, but their chromosomal diversity is remarkable. Their chromosome number varies from the primitive complement of n = 22 to a derived complement of n = 16. Currently 16 extant species are recognised, and there are a further five subspecies of which at least one may ultimately be found to be a true full species.

The following species are currently found in Australia:

Petrogale assimilis **Ramsay, 1877**
allied rock wallaby

Petrogale brachyotis **(Gould, 1841)**
short-eared rock wallaby

Petrogale burbidgei **Kitchener & Sanson, 1978**
monjon

Petrogale coenensis **Eldridge & Close, 1992**
Cape York rock wallaby

Petrogale concinna **Gould, 1842**
nabarlek

Petrogale godmani **Thomas, 1923**
Godman's rock wallaby

Petrogale herberti **Thomas, 1926**
Herbert's rock wallaby

Petrogale inornata **Gould, 1842**
unadorned rock wallaby

Petrogale lateralis **Gould, 1842**
black-footed rock wallaby

Petrogale mareeba **Eldridge & Close, 1992**
Mareeba rock wallaby

Petrogale penicillata **(Gray, 1827)**
brush-tailed rock wallaby

Petrogale persephone **Maynes, 1982**
Proserpine rock wallaby

Petrogale purpureicollis **Le Souef, 1924**
purple-necked rock wallaby

Petrogale rothschildi **Thomas, 1904**
Rothschild's rock wallaby

Petrogale sharmani **Eldridge & Close, 1992**
Mount Claro rock wallaby

Petrogale xanthopus **Gray, 1855**
yellow-footed rock wallaby

Rocky outcrops offer rock wallabies high-quality refugia as well as access to nearby vegetation. Rock wallabies commonly bask in secure sites on these rocky outcrops. In the wet-dry tropics rocky outcrops, particularly sandstone rock complexes, have considerable water run-off, excellent water-holding capacity and significant seepage. Consequently high plant diversity is associated with these rocky outcrops, and hence there is a greater diversity of animals found linked with these than with the adjacent plains country. It is probable that similar, if somewhat less obvious, features are associated with rocky outcrops elsewhere in Australia.

The rocky outcrops where rock wallabies live are effectively 'terrestrial islands' of varying sizes isolated from other similar 'terrestrial island' habitats by a sea of unsuitable habitat, such as grasslands and woodlands. In modern times habitat changes associated with farming and pastoralism have further isolated suitable rocky habitats, with a concurrent lessening of rock wallaby colony viability in many cases.

The rock wallaby's body form is pear shaped, with a lower centre of mass than would otherwise be the case. This increases their locomotor stability when crossing rough and steep habitats. For their body size, rock wallabies have relatively shorter hindlimbs and more robust long bones, as well as highly granulated soles and toe pads to assist them to quickly and safely traverse their rocky habitats. A fringe of stiff hair around the sole may also increase their traction and stability. Their projecting toenails are shorter than those found in other macropods. This is another mechanism to increase stability when traversing irregular and often treacherous terrain. Their tails are long and cylindrical without the broadened base typical of bounding kangaroos. They use their tail extensively to maintain their balance.

Based on chromosomal characteristics reported through the 1950s to 1990s, and more recently supported by molecular studies, there are three groups of rock wallaby.

1. Xanthopus group (*P. persephone* and *P. xanthopus*)
   These two species have near-ancestral chromosome profiles. The Proserpine rock wallaby of north Queensland is most probably the closest extant species to the ancestral rock wallaby. It is found in rocky outcrops and boulder piles with an associated rainforest cover adjacent to open forest (mesic). The yellow-footed rock wallaby is believed to be its xeric equivalent. The yellow-footed rock wallaby has a northern subspecies *P. x. celeris*, found in restricted environments in south-western Queensland, and a more widespread southern subspecies *P. x. xanthopus* that occurs in suitable rocky terrain in South Australia and in a restricted distribution in western New South Wales.
2. Brachyotis group (*P. brachyotis, P. burbidgei* and *P. concinna*)
   This group is an early offshoot in the evolution of the genus that has speciated in isolation in the Kimberley region of Western Australia and the Top End of the Northern Territory. It has been suggested that the nabarlek is better adapted to drier areas than the monjon and short-eared rock wallaby.
3. Lateralis-Penicillata group (*P. assimilis, P. coenensis, P. godmani, P. herberti, P. inornata, P. lateralis, P. mareeba, P. penicillata, P. purpureicollis, P. rothschildi* and *P. sharmani*)
   This group of 11 species has two complexes, the first being the lateralis complex of three species: the black-footed rock wallaby, the purple-necked rock wallaby and Rothschild's rock wallaby found to the west of the Great Dividing Range. The second is the penicillata complex of the remaining eight species, which are found near the eastern seaboard, and in some cases extending inland into the adjacent hinterland. Morphologically species within this complex are remarkably alike.

Rock wallaby morphometrics

Species	Weight kg	Head-body length mm	Tail length mm
Proserpine	3.5–10.2	495–640	515–710
Yellow-footed	6.0–12.0	480–650	565–700
Short-eared	2.2–5.6	405–550	320–550
Monjon	0.9–1.4	305–350	265–290
Nabarlek	1.1–1.7	290–365	220–335
Black-footed	3.1–5.3	450–530	410–605
Purple-necked	2.7–7.1	390–610	445–580
Brush-tailed	4.9–11.0	510–590	500–700
Allied	4.3–4.7	445–590	409–550
Cape York	4.0–5.0	440–565	470–540
Godman's	3.5–5.9	495–640	480–640
Herbert's	3.7–6.7	470–615	510–660
Unadorned	3.1–5.6	455–570	430–640
Mareeba	3.1–5.1	425–550	415–530
Rothschild's	3.7–6.6	470–590	540–705
Mount Claro	3.4–4.7	455–530	435–535

XANTHOPUS GROUP

P. persephone and *P. xanthopus*

Proserpine rock wallaby *Petrogale persephone*

They are believed to be the most primitive of all rock wallabies, and possibly the ancestor of modern rock wallabies. Surprisingly, this species only became known to science in 1976, when they were discovered in the Proserpine region of central coastal Queensland.

They are the largest of the rock wallabies. Their colonies are in rocky terrain within notophyll-microphyll vine forests adjacent to open forest. They browse within the semi-deciduous vine thickets for leaves, fruit and flowers. They regularly move to nearby open areas to graze on both native and introduced pastures. Individuals have home ranges of about 20 ha.

This is probably the rarest of the rock wallabies. According to the 2011 IUCN Red List of Threatened

Figure 1.69: Proserpine rock wallaby browsing fallen leaves off the forest floor. These wallabies stay within the confines of their vine forests for much of the year. When food availability within the forests declines, they venture onto nearby road verges and pastures to forage. (Photo: Hans and Judy Beste.)

Figure 1.70: Proserpine rock wallaby on top of a large boulder. These are typical observation posts for these rock wallabies. (Photo: Hans and Judy Beste.)

Species, they are endangered because their population has a downward trend, there are low numbers and they are restricted to a very small geographical area (about 14 500 ha) of central coastal Queensland. Here the annual rainfall is about 1500 mm at Airlie Beach and 1600 mm on nearby Hayman Island. Winter is by far the coolest period, with mean daily maxima in the vicinity of 23–25°C, compared with 30°C over the summer. Most young are born in the wet season (January–March).

In response to their low numbers and restricted environment, the Queensland Government has mounted a successful Recovery Plan for this species.

Why are they endangered?

- There is a gradual loss of appropriate habitat.
- More and more of their habitat is now under development for human habitation and farms, which has bought the rock wallabies into closer contact with humans and their pets, especially dogs and cats. Road accidents and dog harassment have had major detrimental effects on the population.
- The increased prevalence of toxoplasmosis, a disease carried by domestic cats. Proserpine rock wallabies are particularly sensitive to this protozoan disease, and they have a high mortality rate when infected.
- Hybridisation with the sympatric unadorned rock wallaby.

Threatening factors

- Loss of habitat because of use by man for farming and houses.
- Toxoplasmosis originating from cats.
- Conflict with humans.
- Limited genetic diversity and inbreeding depression.
- Harassment by dogs.

Management action

The Queensland Government has had a successful Recovery Plan in place for several years. Details of this are in the chapter on conservation under Case Study 2.

Yellow-footed rock wallaby
Petrogale xanthopus

Currently they occur across much of their historic geographic range; however, the number of viable colonies within this range has dropped dramatically. The colonies are restricted to arid zones where there are suitable rocky habitats such as boulder piles and the edges of rock faces. Individual animals have home ranges varying from 24 to 200 ha, depending upon the quality of their habitat. Some home range overlap does occur. Surprisingly, it is uncommon for these rock wallabies to use hollows, crevasses or caves to avoid the heat of the day or predators (foxes, wedge-tailed eagles). Instead they prefer lying in the shade of trees during the day. In some cases their daytime refugia may be many kilometres from where they forage and drink.

Yellow-footed rock wallabies are characterised by a greyish-buff back, golden-brown limbs, distinct whitish cheek and hip stripes, as well as a long tail that is orange-brown at its base and darker towards the tip. The tail is long, cylindrical, has a blunt end and frequently has alternating light and dark stripes. There are two distinct subspecies; a dark form, *P. xanthopus xanthopus*, and a lighter coloured form, *P. xanthopus celeris*. There are many colonies of *Petrogale x. xanthopus* in north-eastern South Australia as well as a small population (about 100 animals) about 150 km east-north-east of Broken Hill in New South Wales. *Petrogale x. celeris* is found only in the Adavale Basin of Queensland.

They forage on plateaus above their refugia, as well as among the rocks and boulders of the cliffs that each colony retreats to during the heat of the day.

The bulk of their diet is grass, but some forbs and stellate trichromes (*Ptilotus*, *Solanum*, *Sida* and *Abutilon* spp.) are eaten. As the nutritional value of grass declines over the hot dry periods consumption declines, and browse, notably fallen leaves of woody shrubs, especially of acacias, becomes more important. Even native cypress pines (*Callitris* spp.) are browsed, especially when acacia is in short supply. These wallabies appear to need to drink each day.

Figure 1.71 A pair of yellow-footed rock wallabies grooming each other. Such mutual grooming enhances pair bonding. (Photo: Jiri Lochman.)

Figure 1.72: Yellow-footed rock wallaby among boulders. Note the elongate, well-haired, ringed tail. (Photo: Jiri Lochman.)

Figure 1.73: Yellow-footed rock wallaby bounding down slope. Note its relatively short feet, an adaptation lowering the possibility of them being tripped up on these steep rocky slopes. (Photo: Jiri Lochman.)

BRACHYOTIS GROUP

P. brachyotis, *P. burbidgei* and *P. concinna*

Members of this group are the smallest of the rock wallabies. They are all nocturnal, relatively timid, and forage from the base of their chosen rocky sanctuaries out onto nearby open grasslands. They live in the tropics where there is a distinct wet season, generally December to May, with most rain falling between January and March. June to November is usually very dry.

The effect of body size on adult wallabies in the brachyotis group could be significant. The body weights of monjons range from 0.9 to 1.4 kg, that of nabarleks from 1.1 to 1.7 kg and that of short-eared rock wallabies from 2.2 to 5.6 kg. If one considers representative large males of each species, there is little difference between the monjons and nabarleks, both weighing about 1.5 kg, but the short-ears are more than treble their size at 5.6 kg.

Bettongs and potoroos that have body weights similar to the monjon and nabarlek have high-quality diets and are usually reliant on the high nutritive value of fungi for their well-being. The metabolic needs of the monjons and nabarleks would be similar to those of the bettongs and potoroos. Consequently one would expect that, like them, they could be reliant on the fruiting bodies of hypogeous fungi; however, this does not appear to be the case. The selection of high-quality browse, including flowers, new shoots and seeds supplemented by grasses, meets their nutritional needs. Grasses are more prominent in their diet in the wet season when their nutritional value is highest. Physiologically small animals (less than 1.5 kg), especially lactating females, must be at the limit of microbial fermentation being able to support their high metabolic needs.

Short-eared rock wallaby *Petrogale brachyotis*

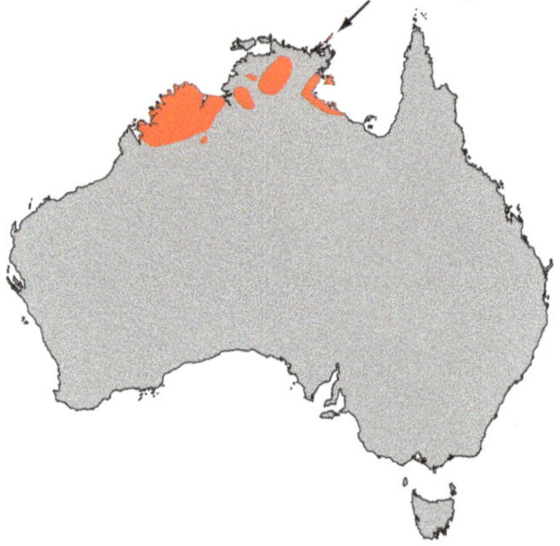

These small inquisitive wallabies are widely distributed from the south-western Kimberley across to the Northern Territory–Queensland border.

Animals observed on and around rocky outcrops in eastern Kakadu National Park remain well hidden over most of the day, particularly over the hottest period. In the early evenings and early mornings they forage around the base of their outcrop and for up to 100 m out onto the adjacent plain. Grasses are a significant dietary component during the wet season,

Figure 1.74: Short-eared rock wallaby sunning itself on rocks at Mount Hart Station in the west Kimberley. These wallabies are readily observed sunning themselves in the early morning and late afternoon. (Photo: Jiri Lochman.)

Figure 1.75: Short-eared rock wallaby on the forest floor in Litchfield National Park. Note the terminal hairy tail tuft. (Photo: Jiri Lochman.)

possibly because of their relatively high nutrient level when fresh and green compared with later in the dry season, when the nutrient level of the grasses drops dramatically as the grasses dry off. Over the dry season browse provides higher quality nutrients. Short-eared rock wallabies are flexible in their diet, ranging from extensive use of *Terminalia latipes* leaves and seeds over much of the year through to browsing on spinifex seeds during the dry season. Short-eared rock wallabies survive the effects of bushfires by browsing the unburnt vegetation of their rock piles.

Short-eared rock wallabies rarely move far from the safety of their home range. They regularly bask in the sun until 10 am and again after 4 pm in shadowy areas of rock faces. They will allow humans to approach them quietly on foot, but they always position themselves to have good escape routes, such as vertical rock faces nearby. They breed throughout most of the year, as evidenced by females having pouch young most of the time.

Monjon *Petrogale burbidgei*

Figure 1.77: Monjon browsing fallen leaves. Note its beautiful tufted tail. (Photo: Jiri Lochman.)

The monjon is the smallest of all rock wallabies. It is restricted to the high-rainfall region of the coastal western Kimberley region and some of the adjacent islands off Western Australia.

Very little is known of the biology of this species. Monjons have been observed to move around their broken sandstone strongholds in the daytime on Bigge Island where there are few predators. On Bigge Island they utilise all potential habitats: rainforest, sandstone escarpments, creek lines, dolerite woodland, *Eucalyptus mineata* woodland, mangroves and beach dunal areas. On the nearby mainland they only move at dusk and evening. Little is known of their reproductive biology except that they have obvious pouch young between August and October. On the mainland typically they lie up during the day in rocky crevices and move out to their feeding grounds in the early evening. Because they are a very small macropod their metabolic demands will be high, so one would expect them to be selective browsers of high-quality food, including fruits, flowers, nuts, seeds, rhizomes, tubers and arthropods.

Figure 1.76: Monjon on forest floor on the Mitchell Plateau of the Kimberley region. (Photo: Jiri Lochman.)

Nabarlek *Petrogale concinna*

These fast, agile wallabies occur in the western Kimberley region of Western Australia, as well as in the north-west and north-east of the Northern Territory and Groote Eylandt.

They have a characteristic habit of holding their tail curled upwards and over their back as they hop. The tuft of the tail fluffs out when they are hopping. Nabarleks are relatively common on the Mitchell Plateau, and appear to like the King Leopold Sandstone areas.

They like broken but not really vertical surfaces. They spend much of their time on the rocks, and venture up to 100 m out onto adjacent flat land to feed. Nabarleks are unique among the macropods in having more than 4 or 5 molars in each jaw quadrant. They have evolved a unique dentition for a rock wallaby by having their lower tooth row curved upwards (as is seen in the large kangaroo species) so that only two molars on each side occlude with their fellows of the upper arcade. The early loss of their premolars facilitates molar progression. Consequently they always have teeth 'in wear' even though the supernumerary molars later in life are of poor quality.

Although their feeding ecology has yet to be examined in detail, data from a few sites indicates that they primarily consume browse and/or forbs throughout the year. However, a greater proportion of grasses are consumed during the wet season when they have high nutritional values. Indigenous people confirm these observations. Because of their high metabolic rate, nabarleks must select highly nutritious food whenever it is available to build up body condition or for females to adequately support suckling young.

Figure 1.78: Nabarlek peering at the photographer. (Photo: Jiri Lochman.)

LATERALIS-PENICILLATA GROUP

LATERALIS COMPLEX

P. lateralis, *P. purpureicollis* and *P. rothschildi*

Black-footed rock wallaby *Petrogale lateralis*

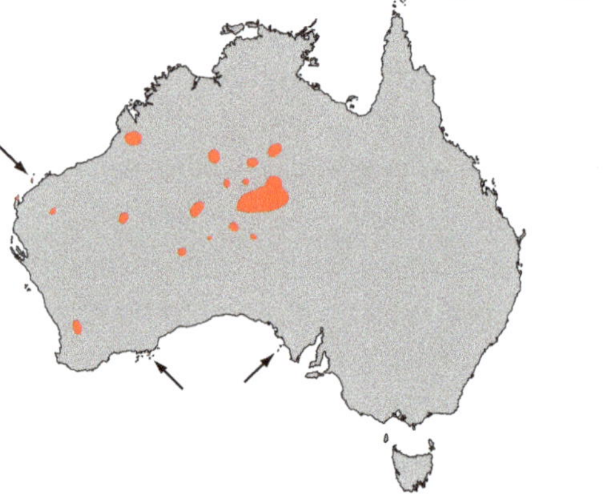

These are commonly called the 'black-flanked' rock wallaby. The species has a disjunct distribution across a geographically widespread area of much of arid Australia, where it has several subspecies and races.

They are rarely sympatric with other rock wallaby taxa; however, in the Pilbara region of Western Australia they are recorded to have hybridised with Rothschild's rock wallabies. Black-footed rock wallabies are large rock wallabies that generally have a dark grey-brown upper coat colour and slightly paler forearms, chest and belly. They have a pale white facial stripe as well as a side-stripe that extends from the axilla to the thigh. They are morphologically and osteologically very similar to Rothschild's rock wallabies.

Over much of their former range their numbers have decreased dramatically. A small relict population (less than 200 animals) on Barrow Island was probably cut off from mainland Australia about 8000 years ago when sea levels rose and formed Barrow Island as well as many other similar islands. Barrow Island rock wallabies became inbred over time and are now genetically depauperate. Many island populations are small and may not be viable in the long term.

Western desert populations, such as those in the Durba Hills–Calvert Ranges, have decreased of late. Predation by foxes and cats of the rock wallabies in these desert areas has restricted them to massive cliffs and deep gorges.

However, the long-term prospects for the rock wallabies in the Durba Hills appears to be good because their rocky habitat is larger and of better quality than in many other desert sites. Smaller rock piles even up to several hectares no longer support these rock wallabies, even though they did so within the timeframe of recent Aboriginal memory. This may be because these rock associations support a rabbit population that is predated upon by foxes. Hence, even in lean times, fox numbers are maintained by predation on rabbits, so predation of rock wallabies, even though low, still occurs. Frequent hot vigorous fires probably disadvantage rock wallabies more than sympatric kangaroo species because of the rock wallaby's inability to recover from the lowered habitat quality; for example, the loss of food resources extending out from the rock piles.

Over the past several decades, widespread reintroductions of these wallabies, coupled with predator suppression, have greatly improved their status. They readily move relatively large distances beyond their rocky refugia to find food. Like most rock wallabies, they forage for high quality grasses and browse.

Figure 1.79: Black-footed rock wallaby preparing to hop up a rocky slope on an island in the Recherche Archipelago. (Photo: Jiri Lochman.)

Figure 1.80: Black-footed rock wallaby with her large pouch young sunning themselves from the safety of a rocky ledge on Barrow Island. (Photo: Jiri Lochman.)

Purple-necked rock wallaby
Petrogale purpureicollis

Until recently they were considered as a subspecies of *P. lateralis*; however, they are now recognised as being a full species.

These animals are found over a large area of the western arid zone of Queensland from Winton in the south-east through to north of the Boodjamulla (Lawn Hill) National Park and across into the Northern Territory. This area is characterised by eucalypt- and acacia-dominated woodlands with an understorey of native grasses, particularly spinifex. The scattered colonies are associated with rough hilly escarpments through to isolated rock piles. These rock wallabies rely on available drinking water sources to survive extended hot periods and long droughts.

They are quite large, nondescript animals, with sandy to grey-brown upper parts and paler underparts. They do not have a distinct side stripe, but they do have a pale facial stripe. Their head and neck may be washed with a purplish tinge. Researchers suspect that the water-soluble cinnabarinic acid or some allied compound may be responsible for the purple face and neck colouring of these rock wallabies. This is most evident in the large alpha males throughout the year.

Rothschild's rock wallaby
Petrogale rothschildi

These large rock wallabies are found only in the Pilbara region of Western Australia. In summer they avoid the high daytime temperatures characteristic of the Pilbara by occupying cool humid caves in the rock piles. These caves reach an average maximal temperature of 30°C and have a relative humidity of more than 75 per cent during the day and more than 95 per cent at night. The species is common across most of its distribution. This is possibly because these wallabies live in extremely inhospitable rocky assemblages where predators are few. However, wallaby numbers have declined along the coastal region, probably because of fox predation, except for the Burrup Peninsula, where 1080 baiting for foxes occurs. Where nearby offshore islands such as Dolphin Island have had all foxes removed, the Rothschild's rock wallaby has bred up into significant numbers.

In inland regions foxes appear not to have colonised the rougher hot terrain; consequently, rock wallaby numbers there are healthy. The popular tourist site of the Millstream-Chichester National Park has a good population of these rock wallabies. Here Pilbara olive pythons *Morelia olivacea barroni* are believed to be significant predators of Rothschild's rock wallabies, especially of the young-at-foot.

Figure 1.81: Rothschild's rock wallaby in natural iron stone environment near Wittenoom in the Hamersley Ranges of Western Australia. (Photo: Jiri Lochman.)

Lateralis-Penicillata group

Penicillata complex

P. assimilis, P. coenensis, P. godmani, P. herberti, P. inornata, P. mareeba, P. penicillata and *P. sharmani*

Brush-tailed rock wallaby *Petrogale penicillata*

These are believed to have been an early offshoot from the ancestral stock of this complex. They have a uniform, shaggy, brown-rufous coat, with their upperparts being darker than their underparts. Their head has a pale cheek stripe and a black dorsal stripe from the eyes to behind the head. Generally their tail is a dark brown-to-black colour with the terminal third being quite bushy. Animals from northern colonies are often paler and have less bushy tails. Colour and pattern variation occurs between and within colonies. The brush-tailed rock wallaby is considered to be near threatened. There appears to be a continuing decline in the number of their colonies over much of their range.

Since Europeans came to Australia this species has been in decline, primarily because of habitat alienation, competition from introduced herbivores and predation, especially by introduced predators.

Currently these wallabies may be found in suitable habitats along the Great Dividing Range from 100-plus km north of Brisbane down to Shoalhaven on the mid coast of New South Wales. At the start of the twenty-first century there were only six known populations within Victoria; all had low numbers (the Grampians population is believed extinct). As for most rock wallabies, little is known of the parameters that make one site suitable for a colony and another apparently similar site unsuitable.

It appears that they prefer mountainous volcanic terrain, notably cliffs and rock piles where caves and rocky ledges abound. Generally such terrain is extremely hard and durable. Colonies generally have steep access slopes, and in most instances steep slopes where the actual core refuge area is located. Connectivity with adjacent potential or real colony sites varies. Like all metapopulations, at any particular time, some colonies can become extinct while others flourish. As long as there are adequate corridors between colonies, migrants can move between colonies, and extinct colonies can be reestablished as others decline. These metapopulations are dynamic but vulnerable entities.

Figure 1.82: Foraging brush-tailed rock wallaby. Note its vibrant colours. (Photo: Jiri Lochman.)

Figure 1.83: A young brush-tailed rock wallaby on an isolated rock. (Photo: Dave Watts.)

These rock wallabies mostly eat grasses and forbs, with browse constituting up to 30 per cent of their diet. Although they are known to be mycophagous, details of the overall contribution of fungi to their diet is not known. They are highly territorial and have home ranges ranging from 2 to 30 ha. Dominant males associate with two to three and even four adult females. Breeding is continuous and there is no apparent seasonality to the births. Like most other rock wallabies, the females are strongly philopatric, while males disperse readily.

The other seven species of the penicillata complex (allied, Cape York, Godman's, Herbert's, unadorned, Mareeba and Mount Claro) bear little morphological similarity to the brush-tailed rock wallaby; however, they are remarkably similar in body form, colour and behaviour to each other. They all have brown to grey upper parts and paler underparts. Their muzzle is usually dark with a pale cheek stripe. Most have a dark armpit area and may have a pale hip stripe. This suite of rock wallaby species has a near contiguous distribution in Queensland from the north-east of Cape York through to Nanango–Kingaroy in the south. Hybridisation has been recorded where any two species are sympatric. The geographic isolation of many colonies, coupled with the complexity of their rocky habitats, make studies and assessment of the frequency of hybridisation difficult.

The allied rock wallaby is the most studied of the complex to date, and one could assume that the findings for this species could be applied broadly to other species in the complex. They have small home ranges during the wet season when good quality food is readily available, but these expand dramatically during the dry season and droughts when food quality and availability are low.

Allied rock wallabies rarely forage far from their rock pile. Forbs and browse constitute most of their diet; however, in the late dry season less palatable species such as *Waltheria* spp. become an increasing part of their daily intake. At no stage do dry grasses form a significant part of their diet. Most mother rock wallabies in this species complex leave their recent out-of-pouch young in a narrow crevice or similar shelter when they move off to feed. Such behaviour could be because such young animals are unable to keep up with the adults as they move to and from the feeding areas. As the very young have a high risk of predation by cats and foxes, such behaviour makes sense; however, it is not a defence against snakes, especially pythons.

With allied rock wallabies only about two-thirds of the behaviourally monogamous pairs were actually genetically monogamous. Extra pair paternities account for genetic diversity within the species. Shortages of shelter sites in colonies may force young rock wallabies into available territory where suboptimal males are established. Intercolony dispersal has been recorded as 2 km for allied rock wallabies. The average lifespan for allied rock wallabies is about 7 years, but individual animals have been recorded to live for 16 years.

DISTRIBUTIONS OF THE QUEENSLAND PENICILLATA COMPLEX

P. coenensis

P. godmani

P. mareeba

P. sharmani

P. assimilis

P. inornata

P. herberti

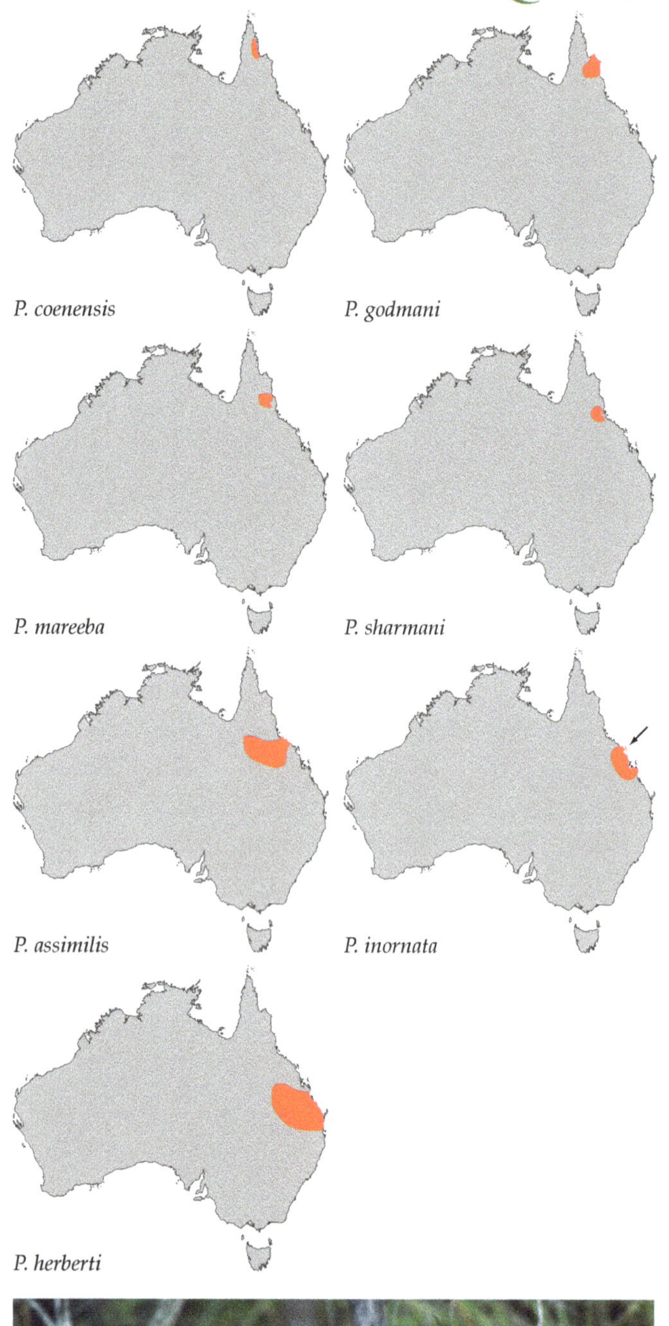

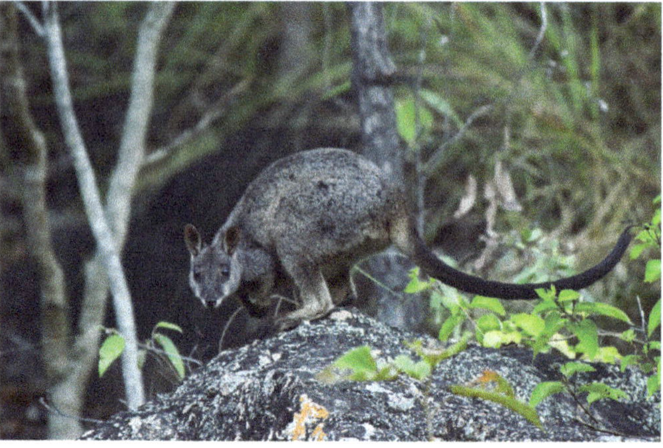

Figure 1.84: Allied rock wallaby ready to flee into nearby forest. (Photo: Jiri Lochman.)

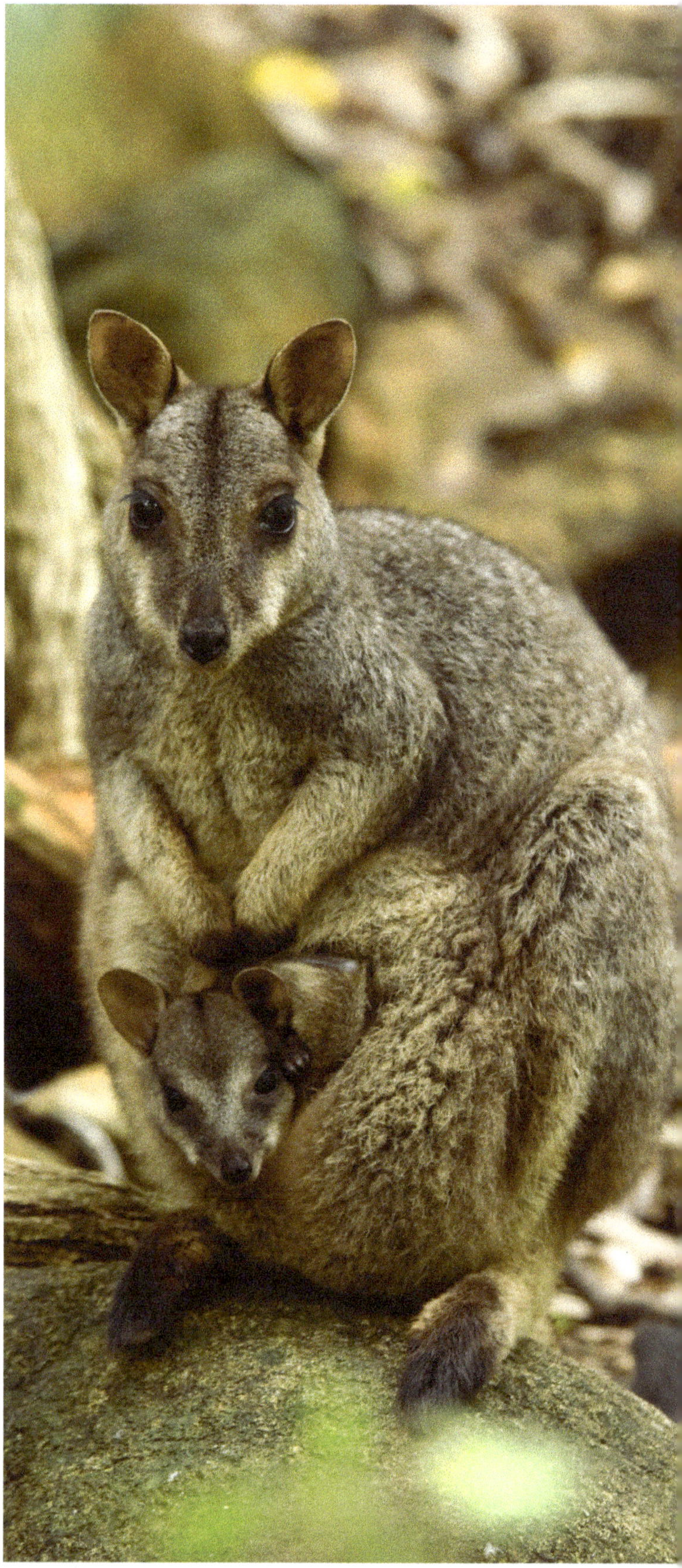

Figure 1.85: A female allied rock wallaby with her large pouch young. (Photo: Jiri Lochman.)

Figure 1.86: A pair of allied rock wallabies at Bowling Green Bay National Park near Townsville. (Photo: Jiri Lochman.)

Figure 1.87: A well-furred Mareeba rock wallaby pouch young peering out of its mother's pouch. (Photo: Jiri Lochman.)

Figure 1.88: A female Mareeba rock wallaby with her dependent pouch young at foot both sitting sun bathing. Note her everted relaxed pouch. (Photo: Dave Watts.)

Although studies of the reproductive biology of the rock wallabies are few, the following is a summary of available data.

	Postpartum oestrus	Oestrus cycle	Embryonic diapause	Gestation period	Pouch life	Pouch exit	Weaning
P. assimilis	Yes	29–34		29–34	180–210		~330
P. brachyotis							
P. burbidgei							
P. coenensis							
P. concinna		32–35		30–32		180	
P. godmani							
P. herberti							
P. inornata	Yes	30–32	Yes	30–32	180–210		
P. lateralis	Yes	30		30	180–210		330
P. mareeba							
P. penicillata			Yes	~30	187–200		270–300
P. persephone	Yes	33–38	Yes	30–34	210		330–360
P. purpureicollis	Yes	36–38		33–35	180–210		270–360
P. rothschildi							
P. sharmani							
P. xanthopus	Yes	32–37		31–33	180–210		

Figure 1.89: Two male Mareeba rock wallabies sparring on a rocky slope. (Photo: Hans and Judy Beste.)

Generic threatening factors for rock wallaby colonies

- Predation by introduced predators, particularly foxes, dogs and in some cases feral cats.
- Loss and fragmentation of habitats.
- Degradation of habitat because of grazing activities of domestic and feral herbivores such as goats, rabbits, sheep, cattle, pigs, horses, donkeys and camels.
- Competition with introduced herbivores.
- Changing fire regimes, which reduce the abundance and diversity of ground forage.
- Increases in tourism, causing disturbances to the colonies as well as habitat.
- Climate change, with its inevitable changes to the plant communities associated with rock wallaby habitats.
- Introduction of novel disease.

Management action

- Where appropriate, undertake regular feral predator control around colony sites.
- Retain rocky habitats intact.
- Ensure grasslands and open forests adjacent to colony sites are managed to provide suitable graze and browse for the rock wallabies.
- Manage domestic herbivores within the foraging area of each colony.
- Manage the use of fire in areas adjacent to rock wallaby colonies.
- Undertake weed control in and around their habitats.
- Retain suitable corridors between colony sites.
- Endeavour to stop any introduction of disease into any population.
- Raise landowner awareness of the presence of colonies on their land and advise them on best possible management practices for these sites.
- Protect all colony sites from human disturbance and interference.
- Manage human development near rock wallaby habitats.
- Where necessary, use translocations and reintroductions into secure and managed areas of suitable habitat. Ensure that ongoing management of these populations is appropriate and well resourced.
- There is a great need for further studies on this diverse group. Knowledge of the basic biology of many species is poor.

Pademelons

Pademelons are small (5–10 kg) macropods, with an ancient lineage that are now found in New Guinea and along the eastern side of mainland Australia through to Tasmania. There are seven species in total, with four species restricted to New Guinea, two to Australia and one, the red-legged pademelon *Thylogale stigmata*, being found in both Australia and New Guinea. Pademelons may have given rise to the rock wallabies in Australia. It is suggested that pademelons reached New Guinea from Australia about 2 mybp. In modern times pademelons are found mostly in heavily forested areas and adjacent grasslands. Unfortunately, the New Guinea members of the genus are all either endangered or vulnerable to extinction. The three Australian species are not at risk.

Red-legged pademelon *Thylogale stigmatica*

These are medium-sized macropods found within the edges of rainforest, dry vine forest, as well as in wet sclerophyll forests. Even with the massive losses of suitable forest habitat, this species is still common from the tip of Cape York in the north through to about Tamworth in northern New South Wales. Red-legged pademelons are also found in south-western Papua.

Figure 1.90: A female red-legged pademelon accompanied by her large dependent young within their rainforest habitat. (Photo: Hans and Judy Beste.)

Figure 1.91: A female red-legged pademelon eating a piece of watermelon. Such tame pademelons are often a tourist attraction. (Photo: Dave Watts.)

They are characterised by a brown-grey grizzled pelage with a bright rufous face, flank and hindlimbs. Like many kangaroos, they have a pale facial stripe and the suggestion of a hip stripe.

They are generally solitary animals, but may form feeding aggregations in the evenings on especially desired habitat such as improved pasture. In the southern portion of their range they are primarily forest denizens, with up to 99 per cent of their diet being dicotyledonous browse. In the far north of Queensland, unlike most kangaroos, this species may be active throughout much of the day.

They have been reported to browse during the day within the rainforest on dicotyledon leaves, including dead leaves off the forest floor and available fruits, such as those of rainforest figs *Ficus* spp. and Burdekin plums *Pleiogynium timorense*. In some instances their browsing of small forest floor plants may impede its regeneration. Their nocturnal range extends to just outside of the forest edge, where they opportunistically graze improved pastures (*Paspalum notatum*, *Cyrtococcum oxyphyllum*). Consequently, in some areas their daily intake may be of roughly equal amounts of dicotyledons and monocotyledons. They are mycophagous, but the extent of this is unknown. Dingoes have been reported to prey opportunistically on pademelons along the forest edges. Home range size is related strongly to habitat type; those of northern rainforests are between 1 and 4 ha.

They exhibit sexual dimorphism, with males weighing up to 6.5 kg and females up to 4.2 kg. Their head-body length ranges from 470–536 mm in males to 386–520 mm in females. Tail length ranges from 372–473 mm in males to 301–445 mm in females. Sexual maturity is reached at 48 weeks in females and 66 weeks in males. There is a postpartum oestrous following birth, with mating occurring within 2–12 hours. Their oestrous cycle is 29–32 days and gestation period is 28–30 days. Pouch exit is between 26 and 28 weeks. Weaning is between 37 and 39 weeks.

Threatening factors

- Predators such as spotted-tailed quolls, amethystine pythons, foxes and dingoes.
- Loss of preferred habitat – dense rainforest with nearby pastureland.

Management action

- Control of foxes and, where appropriate, dingoes.
- Retention of preferred habitats.

Red-necked pademelon *Thylogale thetis*

Figure 1.92: Red-necked pademelon browsing in Barrington Tops National Park. (Photo: Jiri Lochman.)

These are similar in most respects to the red-legged pademelon. They are true denizens of the rainforests and dense eucalypt forests of southern Queensland through to southern New South Wales.

They access their feeding grounds via well-established runways. They are mostly browsers within their forest homes. They rarely move more than 100 m from the forest edge to graze on pastures adjacent to their forest refugia. Red-necked pademelons consume a large number of hypogeous fungal species throughout the year, with highest consumption in autumn and winter. Home ranges vary from 5 to 30 ha. Where they are sympatric with red-legged pademelons, they appear to exclude them from pastures adjacent to the daytime forest refugia.

As their name suggests, these pademelons have a characteristic rufous neck and upper torso. The remainder of their pelage is mostly a grizzled grey-brown. Females weigh up to 4 kg and males to 7 kg. Head-body length averages 420 mm in females and 520 mm in males. Tail length averages 350 mm in females and 430 mm in males.

Females are sexually mature at 17 months and are continuous breeders.

Threatening factor

Loss of preferred habitat.

Management action

While their rainforest core habitats and adjacent pasturelands remain available to these wallabies, the species should be secure.

Figure 1.93: A female red-necked pademelon with her inquisitive pouch young in low-altitude rainforest of Barrington Tops National Park. (Photo: Jiri Lochman.)

Tasmanian pademelon *Thylogale billardierii*

For obvious reasons, they are often called the rufous-bellied pademelon. They are now found only in Tasmania and adjacent islands; however, within the last 200 years they have been common on the adjacent south-eastern mainland of Australia.

They mostly inhabit moist forests and densely vegetated thickets and gullies. They are well adapted to alpine conditions. Their home range may be as large as 170 ha; however, most are considerably smaller. Typically they graze soft grasses and browse forbs and low shrubs. They are particularly fond of flowers. They rarely venture beyond 100 m of their refugia.

Even though they are quite small, commercial harvesting of these pademelons was undertaken, particularly on Flinders Island. In some areas of Tasmania they are still culled where they are pests of commercial crops.

They are sexually dimorphic, with males weighing up to 12 kg and females to 10 kg. Head-body length averages 550 mm in females and to 630 mm in males. Tail length averages 320 mm in females to 417 mm in males. They have dense fur that varies from a dark brown to a grey-brown above. Their underparts are a reddish-brown with a pale whitish chest.

Sexual maturity is between 13 and 17 months. Males have well-established dominance hierarchies. Breeding is continuous throughout the year, with most births being in late autumn and early winter. There is a postpartum oestrous. Their gestation period is approximately 30 days. They have a pouch life of about 200 days.

Management action

The species is common in its island refuge and does not appear to be in need of any special management.

Figure 1.94: A Tasmanian pademelon and her pouch young foraging in a forest opening created by clear felling with subsequent bracken fern invasion in the Asbestos Range of Tasmania. (Photo: Jiri Lochman.)

Figure 1.95: A Tasmanian pademelon and her pouch young both browsing a flowering native heath. (Photo: Dave Watts.)

Quokka *Setonix brachyurus*

This is a monospecific genus restricted to a few mainland sites and offshore islands of southern Western Australia. They are short, squat, grizzled, grey-brown wallabies that have an unusually short, bare tail. An average male weighs 3.6 kg and has a head-body length of 487 mm and tail of 289 mm. An average female weighs 2.9 kg and has a head-to-body length of 468 mm and a tail length of 265 mm. They have short rounded ears. Latitudinal effects are significant, with Rottnest Island (32°00′S) animals being much larger than Bald Island (34°55′S) animals.

Mainland animals are genetically the most representative of the species. Even though they occur at more than 25 mainland sites, individual colony sizes are usually small. Most sites are either in swampy regions or coastal heath. Many inland sites are in jarrah forests characterised by dense thickets, especially those alongside streams having thickets of tea-tree *Taxandria linearifolia* and sword grass *Lepidosperma effusum* and *L. tetraquetum*. Wherever they are found, they occur in colonies that create and use distinctive runways (150 mm high by 100 mm wide) to ease their passage through the undergrowth. Quokkas need a mosaic of seral stages in an area for them to persist. Recently burnt areas (less than 10 years) provide a diversity of food sources, while areas unburnt for up to about 25 years provide dense refugia from predation. Older unburnt areas (more than 25 years) on their own appear unable to sustain a quokka population.

The largest number of quokkas occurs on Rottnest Island near Perth, where the population estimates vary from 8000 to 12 000 individuals. Rottnest Island lost its original dense tree cover of *Callitris*, *Acacia* and *Melaleuca* shrubs because of bushfires and firewood collecting in the 1800s. What remains is a particularly harsh habitat covered by low tussock grasses interspersed with clumps of native and introduced shrubs. Another robust, but considerably smaller population (200–600) is found on Bald Island near Albany.

Quokkas prefer new young growth, and are well adapted for browsing leaves and stems. Mainland populations feed on a wide variety of dicotyledonous shrubs, notably *Thomasia* spp., *Dampiera hederacea*, *Mirbelia dilatata* and peppermint *Agonis linearifolia*. Quokkas eat small amounts of fungi. Where browse is limited they survive on grasses, but in some seasons, such as on Rottnest Island by the end of summer, quokkas have only poor quality food. At this time they subsist on succulents, grasses, sedges and shrubs, where the available vitamin E, an antioxidant that combats membrane oxidation, is low. Consequently most animals have a poor body condition. Following the first rains of autumn there is a flush in plant growth. However, this is high in polyunsaturated fats that cause excessive oxidation of cell membranes throughout the body. Consequently many animals develop muscular dystrophy and cardiomyopathy and some die. The early administration of vitamin E protects against this condition, and in clinical situations it may save a moribund animal. Quokkas have a relatively high requirement for water, necessitating their living in close proximity to fresh-water sources throughout the year.

In the wild, sexual maturity is 8–9 months in females and about 13 months in males. On Rottnest Island they have a restricted breeding season, with most young being born in February and March. This is probably because over the summer and autumn period the available food is of extremely low quality, so consequently the females become anoestrous. However, on the mainland they are continuous breeders, a feature linked to the availability of a more favourable food resource base throughout the year. They have a postpartum mating within 1 day of the young being born. Oestrous is 28 days, and the gestation period between 26 and 28 days. They have embryonic diapause. Pouch life is about 24–28 weeks, but suckling may continue for an additional 8 weeks.

Their conservation status is vulnerable, according to the Commonwealth *Environmental Protection and Biodiversity Conservation Act* 1999; and according to the Western Australian *Wildlife Conservation Act* 1950 they are regarded as 'fauna which is rare or likely to become extinct'. The 2011 IUCN Red

Figure 1.96: A photogenic quokka on Rottnest Island standing among dried 'pussytails', an introduced African weed. (Photo: Jiri Lochman.)

Figure 1.97: Two mainland quokkas feeding. (Photo: Jiri Lochman.)

List of Threatened Species classifies the quokka as vulnerable.

Threatening factors

- Predation, mainly by foxes, and to lesser extent feral cats, is of ongoing significance. Any diminution in fox-baiting programs within the quokka's current distribution would seriously threaten the ongoing viability of the small mainland colonies.
- Habitat loss and fragmentation, which ranges from ongoing logging operations to agricultural and urban development. All have direct impact by lessening the area of suitable habitat, especially swamps, as well as more subtle changes, such as increasing distances from daytime refugia to suitable feeding sites. Ongoing activities by large numbers of feral pigs in the riverine edges and swamps have caused serious degradation of some previously suitable habitats.
- The high frequency of fire (including controlled burns) reduces the mosaic of seral stages essential for the ongoing viability of many habitats to support quokka colonies.
- Negative interaction with human activities: for mainland populations this ranges from simple disturbance caused by recreational activities, such as camping in areas adjacent to quokka colonies, through to the occasional road kill. The latter can be particularly important, given the decline in population size of many colonies.
- There is a lower than ideal level of genetic variability in the entire quokka population. It is reported that quokkas have had restricted gene flow across the species' range since the

late Pleistocene epoch: 1.8 mybp–11 000 years ago. With island populations this is simply a result of long-term isolation with its inherent inbreeding. With the mainland population, the ravages of environmental changes, especially throughout the twentieth century, resulted in the extensive metapopulation network being severely disrupted. Currently the remnant population consists of about 25 separate small colonies where genetic diversity is low. Even so, the mainland population has greater genetic variability than do the island populations.

- Novel disease is considered to be a possible catastrophic factor to any naïve population of marsupials. Although no specific disease entities can be implicated as disease risks of quokkas, most populations are naïve. Hence the possible risk cannot be ignored.

Management action

- Maintain predator control and, where necessary, pig control in and around all colonies.
- Continue ongoing habitat management and, in particular, ensure burning programs create a mosaic of seral stages to enhance the habitat available to all quokka colonies.
- Minimise the effects of human–quokka interactions.
- Continue monitoring of all populations.
- Limit threatening factors such as the introduction of disease to animals on their offshore island refugia.
- Use reintroduction programs to suitable mainland habitats.
- Continue research into their biology, ecology and genetic makeup.

Figure 1.98: A female mainland quokka with her large dependent young browsing on the forest floor. (Photo: Jiri Lochman.)

Swamp wallaby *Wallabia bicolor*

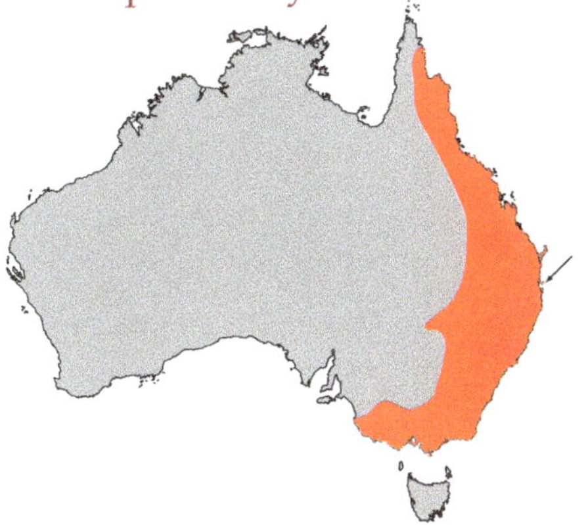

The swamp wallaby, commonly called 'black wallaby', 'black-tailed wallaby' and 'stinker', is classified in a mono-specific genus. This common species has a distribution along much of the east coast of Australia, from Cape York in the far north through to Mt Gambier in southern South Australia. They have a dark brown to black, shaggy coat, including the tail. Their underparts are much lighter, being creamy to orange. Their armpits are black. They frequently have a pale cream to off-white facial stripe that runs from the lower muzzle to below the ear.

Figure 1.99: A vigilant swamp wallaby with ears pricked. (Photo: Raoul Slater.)

Males weigh up to 20.5 kg and females to 15.4 kg. Male head-body length is up to 860 mm and 750 mm in females. Tail length is up to 860 mm in males and 730 mm in females. The South and North Stradbroke Island populations (arrowed on map) of swamp wallabies are about 20 per cent smaller than the nearby mainland animals at the same latitude.

This is a solitary species that lies up during the day in dense thickets in a variety of forest types such as brigalow and open eucalypt woodlands. They are unusual among the macropodids because they mark their territories with faeces. They are well adapted to moving at speed through thick undergrowth because they hop with their head and shoulders held low and their tail extended straight behind. They have a diurnal tendency. They are specialist browsers, having a dentition where the incisors allow precise selection for plucking or shearing of browse. More sturdy food items are sheared off using their sectorial premolars.

Their elongate premolar is longer than the first molar. The molar arcade is horizontal and has lophs adapted for fine shearing and basic grinding of the leaves (herbs, shrubs, ferns and grasses) that comprise much of their diet. They feed on fruiting bodies of a variety hypogeous fungal species, particularly *Mesophellia glauca*, and consequently are an important dispersal agent for their spores. This is of special importance immediately after intense fire when the surface fungi, including spores, are destroyed. Their activities facilitate the recovery of mycorrhizal-forming plants.

In the wild, sexual maturity is at about 15–18 months in females and in males. They appear to be continuous breeders in the northern part of their range, but more seasonal in southern latitudes, where most young are born in winter. They have a postpartum oestrus, and are unusual among kangaroos in having a gestation period (33–38 days) that is longer than their oestrus cycle. Delayed implantation is usual. Pouch life is for about 8–9 months; suckling may continue until 16 months. This wallaby is common over a wide geographic range and, as such, no special measures are needed to ensure its survival.

Figure 1.100: Swamp wallaby in scrub on Mount Kaputar National Park near Narrabri in New South Wales. (Photo: Jiri Lochman.)

2. Adaptations and function

Evolution of the kangaroo body form

The unique kangaroo body form is seen in all Macropodiformes with the exception of the musky rat kangaroo, the most primitive of extant kangaroos. Instead, this has hindlimbs that are similar in size to its forelimbs. Musky rat kangaroos are strictly quadrupedal and have a synchronous plantigrade gait that is used for both slow and fast locomotion as they move through their rainforest habitat.

This suggests that what we know as the kangaroo body form evolved primarily because of the acquisition of a suite of musculoskeletal adaptations for bipedal hopping (saltation). Within the kangaroo's ancestors the transition from a quadrupedal gait to saltatory gait is believed to have occurred during the mid Eocene or early Oligocene epoch, 45–30 mybp. The development of elongated hindlimbs, with most of their muscle mass in the upper leg and pelvis, was the most significant change. Simultaneously

Figure 2.1: A musky rat kangaroo with the remains of a green katydid that it was holding in its paws while eating it. The back legs of the musky rat kangaroo are only slightly larger than the front legs. (Photo: Stan Breeden.)

Figure 2.2: Western grey kangaroo using a slow pentadactyl crawl. It is taking its body weight on its arms and tail before lifting its legs forward. (Photo: Dave Watts.)

there was a reduction in the size of their forelimbs as their importance for locomotion lessened. Terrestrial kangaroos have more than 50 per cent of their total body muscle mass in their hindlimbs, about 22 per cent in the back extensors, less than 5 per cent in the forelimbs and about 5-plus per cent in the tail. The macropodid body form was also influenced by the development of a dentition that allowed the efficient acquisition and processing of plant material coupled with a very large specialised elongate forestomach for the microbial fermentation of plant material. The remaining body systems had little influence on the general body form of macropods.

Adaptations for hopping and climbing

At low speeds all gaits are energetically expensive. For slow terrestrial progression macropodids use either a synchronous quadrupedal crawl or a pentapedal progression that incorporates their tail as a prop support during a synchronous four-footed crawl. The pentapedal crawl has a quadrupedal support phase alternating with a phase of support by the forelimbs and the tail while the rear legs are suspended.

Among mammals there are two basic patterns of hopping: the first, as seen in rabbits, is quadrupedal hopping, where the forelimbs and the hindlimbs contact the ground alternately. The second is bipedal hopping, best seen in kangaroos, where only the hindlimbs are used.

Terrestrial kangaroos

Bipedal hopping is a highly efficient gait for both speed and long-distance travel. It has needed many morphological and physiological adaptations, and these have resulted in a body form that is virtually an elongated cone. Here the large hind legs and tail together with the abdomen form the base. A small tapered chest region having small forelimbs surmounts this. The head and neck form the apex. During the take-off phase of bipedal hopping there are large propulsive forces that tend to cause the body to twist; consequently the kangaroo's centre of mass is lowered towards the sacrum, and the vertebral column is strengthened. The vertebral column strengthening has been by shortening of the trunk and modification of the vertebral column, its supporting ligaments and associated musculature.

Bipedal hoppers are characterised by large feet, a long tail and small short forelimbs. Their basic hop consists of an initial propulsion phase, then a flight phase followed by a landing phase. During the propulsion phase there is strong contraction of the hip and hock (ankle) extensor muscles. The legs are extended backwards and simultaneously the tail moves downward closer to the body. During flight the legs are moved forward in anticipation of landing while the tail extends further backwards. On landing all components of both legs absorb impact forces, and in particular the large tendons store some as elastic energy. Some of the impact forces are transferred from the legs into the vertebral column. Hopping in the larger species such as the red kangaroo is accompanied by the head moving up and down through an angle of about 10 degrees to counterbalance the backward swing of the body during the thrust of take-off (head down when legs go back). Bipedal hopping is a stable gait, particularly in the larger grazing species. Even though tree kangaroos have redistributed much of their lower body mass to their upper body to facilitate their climbing, they still use bipedal hopping when on the ground.

The form of the macropodid axial skeleton is typical of terrestrial mammals. The vertebral column has 7 cervical, 13 thoracic, 6 lumbar, 2 sacral and 21–25 caudal vertebrae. Their vertebrae are opisthocoelous: where the vertebral body is convex at the front and concave behind. Except for the distal caudal vertebrae, each vertebra interlocks cranially with preceding and caudally with succeeding vertebrae. Among the larger bounding kangaroos there has been a significant enlargement of the articular processes to unite the lumbar vertebrae firmly. There has also been the development of large dorsolateral extensions of the transverse processes of the lumbar vertebrae. These lumbar extensions provide extensive attachments for the hypaxial and epaxial musculature. Similarly, caudal vertebrae 1–6 (the first six tail bones) have well-developed processes that allow for the complex muscle and tendinous arrangements of the tail. In the larger species the first caudal vertebra has a large articulation with the sacrum that allows the tail to swivel during pentapedal crawling. Further distally the caudal vertebrae become more and more simplified. Except for caudal vertebrae 1–3, haemal arches lie ventrally and transversely across sequential intervertebral discs for most of the length of the tail.

Figure 2.3: Western grey kangaroo skeleton. The most remarkable features of this skeleton are its elongate tibia and specialised foot, the large straight pelvis and epipubic bones. In addition, its lumbar and tail vertebrae are sturdily developed to accommodate large muscle attachments. (Photo: Joe Hong.)

The vertebral column of the large grazing kangaroos has evolved to accommodate distribution of forces acting on it during hopping. There is the dramatic increase in the size of the intervertebral discs from the thoracic vertebrae (at chest level) to those at the lumbosacral junction (where the vertebrae join the pelvic bone). In a medium-sized western grey kangaroo, the intervertebral disc depth at the junction of the last thoracic and first lumbar vertebrae is about 4 mm, while at the lumbosacral junction it is about 17 mm. This remarkable size gradation reflects the distribution of forces acting on the vertebral column during hopping.

Quite distinct from the larger members of the macropodiformes, the smaller members, such as the potoroidae and the quokka, do not bound. When they are hopping they have either a crouched-over stance or a semi-upright stance. These smaller members are highly agile, and can change direction much more rapidly than their much larger counterparts. The smaller species are characterised by relatively shorter, less muscular hindquarters; lumbar and caudal vertebrae without enlarged mammillary or transverse processes; and simpler intervertebral connections.

The sexual dimorphism in terms of size in an adult family group of red or grey kangaroos is dramatic – the adult males are double to triple the body mass of the adult females. Because the large males continue to grow slowly for most of their lives, an old dominant male can reach an impressive body size.

The dominant males have superbly developed chest muscles as well as proportionately larger forearms and hands than their females. In most kangaroo species, arm length is a clear indicator of the sexes. In males it may be up to 12 per cent longer than in females. Within groups of males, individuals with longer arm lengths are more likely to be dominant, because in aggressive interactions males with longer and stronger arms are most likely to win an encounter and consequently have more matings.

Figure 2.4: Quokka skeleton. In most respects its skeleton is very similar to that of the larger grazing macropods. Because its fast hopping is a short proppy hop its lower vertebral column is simpler than that of its larger cousins. (Photo: Joe Hong.)

In macropods, the appendicular skeleton is highly modified for bipedal locomotion. The forelimb is relatively simple in form. The scapula (shoulder blade) is unremarkable except for its possession of an elongate protuberant acromial process for articulation with the clavicle (collar bone). Kangaroos have a pronounced clavicle, which is a rigid strut running between the manubrium (top of the sternum) and the scapula. The clavicle is held off the chest wall to maintain optimal muscle length and strength of associated muscles, especially those operating above the horizontal plane. This is particularly important for climbing tree kangaroos. It also acts as a support suspending the scapula and forelimb. It enables the shoulder joint to move laterally through 180°, thus increasing its range of movement. This in turn allows a greater range of movement and flexibility of the hands. Distally the humerus has highly expanded lateral and medial bony flanges for muscle attachment. The radius and ulna are elongate and slender, except in the tree kangaroos, where they are robust: an adaptation for climbing. The circular cross-sectional profile of the proximal radius facilitates rotation of the forearm. This is essential for tree kangaroos to be able to grasp and hold a tree trunk when they are climbing.

The manus (hand) has five simple digits that are not opposable. The outer and inner digits are about two-thirds of the dimensions of the middle, largest digit. The fourth and second digits are intermediate in size. Each digit is tipped with an elongate, laterally compressed, down-curved claw. The claws are tough, and extend well beyond the terminal segments of the digits. The claws of potoroines are particularly long and robust because they are used extensively when digging for truffles and tubers. Browsers use their forelimb dexterity to manipulate food into their mouth.

The elongate hindlimb is a striking morphological adaptation, as all the weight-bearing long bones are significantly longer than those of eutherians (non-marsupial mammals) of similar body mass. The hindlimb long bones are much longer than those of the forelimb. Surprisingly, the femur, tibia, fibula and pes are all relatively slender. In macropods the locomotor forces associated with hopping are mostly compressive and are directed along the main axis of the hindlimb. The principal axis for locomotory forces in the hindlimb is through the talus and tibia. The tibiotarsal articulation is a confined strong hinge joint with little lateral movement. The distal hindlimb of a macropod is extremely elongate and has a narrow profile. The plantar surface is characteristically long and narrow. It lacks a first digit; the second and third digits are both extremely narrow and elongate. Soft tissues bind digits 2 and 3 together for most

Figure 2.5: Red kangaroos. The male, on the right, is much larger than its female. Males may be twice the body weight of females. Males have large broad chest muscles and highly developed upper arm muscles. (Photo: Stuart Miller.)

of their length, a condition called syndactyly. The small separate claws are used together as grooming implements. The fourth digit is extremely long and robust. The outermost digit, the fifth, is quite large.

The fourth digit is the primary propulsive skeletal unit of the foot and is critical in the dissipation of compressive forces of landing as well as in general body support during slow locomotion and standing. The fifth digit provides functional support to the fourth digit. Because of the extreme demands of traversing their environments both rock wallabies and tree kangaroos differ from most other macropods by having shorter and relatively more robust feet. They also have granulated footpads that lessen the possibility of their slipping.

The pelvis of macropodids is relatively large and elongated, with all elements being rather slender. The large muscle mass originating and inserting into the pelvis, femur and vertebral column essential for bipedal locomotion keeps their centre of mass low, thus facilitating effective and efficient hopping.

The energetics of quadrupedal versus saltatory locomotion is complex. A simple explanatory example is to consider locomotion in a frog. When walking, a frog uses less energy per stride than it would use in a jump, but the overall length of a jump compensates. For a given distance, jumping is nearly twice as energetically cost effective as walking. Similarly, by adopting a bipedal hopping gait a macropod can move at a more rapid rate without dramatically increasing energy consumption. In the kangaroo species with a body mass of less than 5 kg, the faster the animal hops, the greater the energy use. At slow speeds (<15 km h^{-1}), when body mass is taken into consideration, quadrupedal locomotion is more energetically efficient than saltatory locomotion, but at median speeds saltatory locomotion has lower energetic costs.

However, in species with body weights above 5 kg – mostly wallabies and grazing kangaroos – the energy use is high at low speeds, but at intermediate speeds the utilisation of energy levels off only to increase again at high speeds. Professor Terry Dawson of the University of New South Wales, who studied red kangaroo locomotion for several decades, found that when they are hopping at low (<15 km h^{-1}) and fast (>35 km h^{-1}) speeds, it is energetically expensive. However, hopping at speeds between 15 and 35 km h^{-1} is highly efficient because of the energy saved, partly

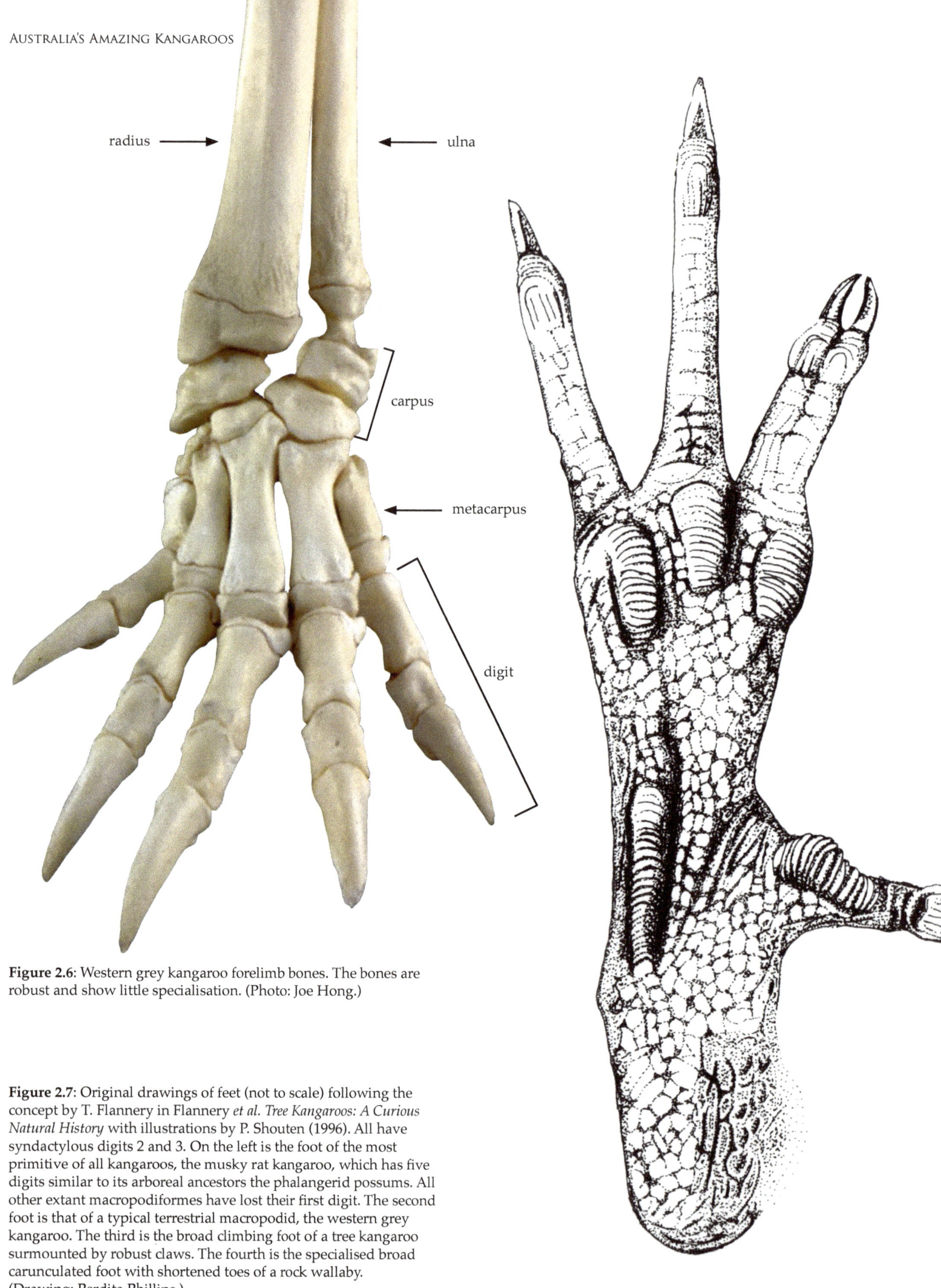

Figure 2.6: Western grey kangaroo forelimb bones. The bones are robust and show little specialisation. (Photo: Joe Hong.)

Figure 2.7: Original drawings of feet (not to scale) following the concept by T. Flannery in Flannery *et al. Tree Kangaroos: A Curious Natural History* with illustrations by P. Shouten (1996). All have syndactylous digits 2 and 3. On the left is the foot of the most primitive of all kangaroos, the musky rat kangaroo, which has five digits similar to its arboreal ancestors the phalangerid possums. All other extant macropodiformes have lost their first digit. The second foot is that of a typical terrestrial macropodid, the western grey kangaroo. The third is the broad climbing foot of a tree kangaroo surmounted by robust claws. The fourth is the specialised broad carunculated foot with shortened toes of a rock wallaby. (Drawing: Perdita Phillips.)

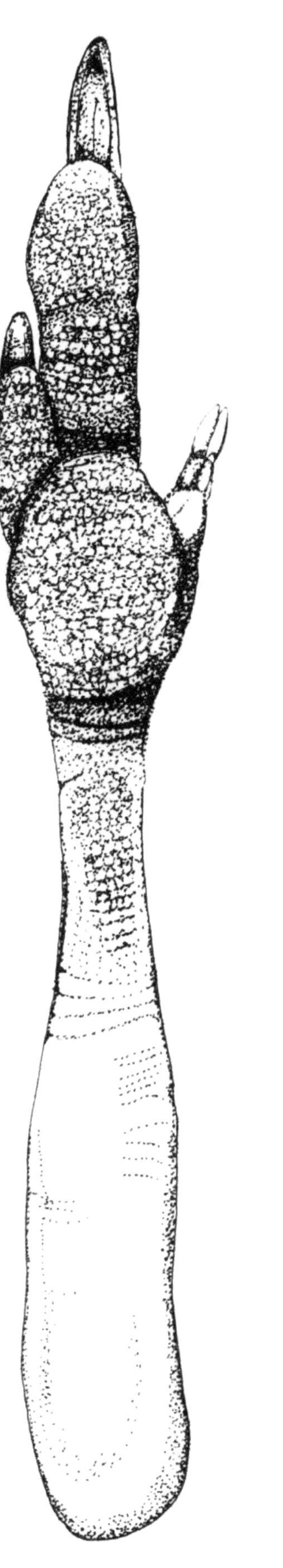

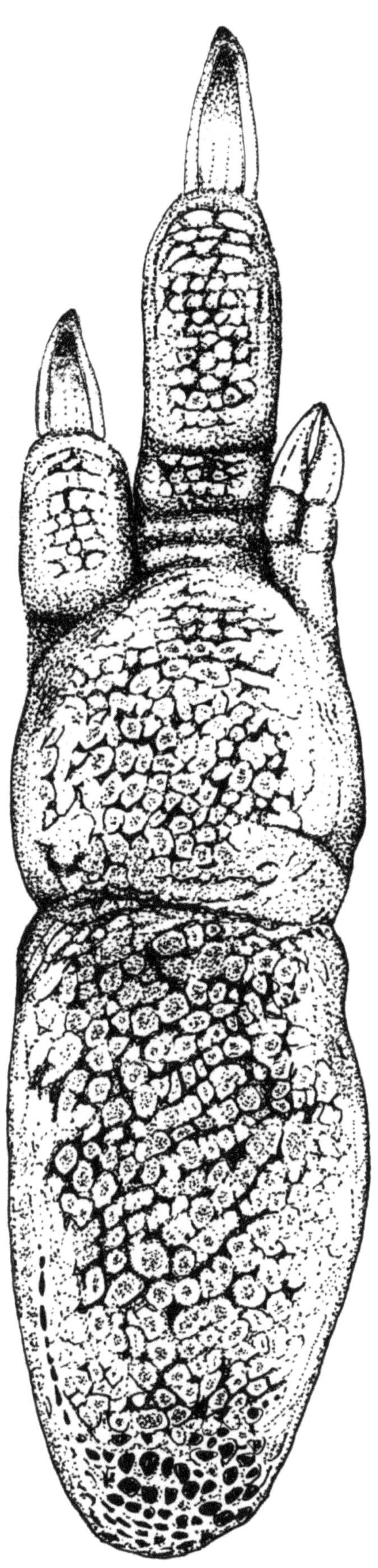

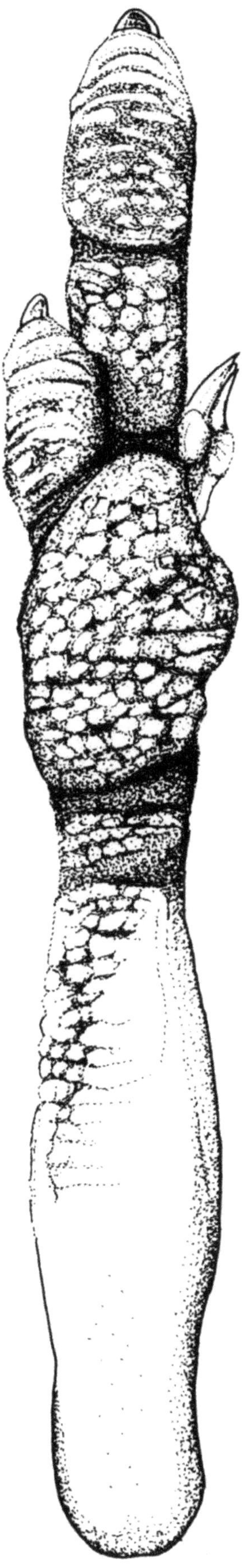

Figure 2.8: Eastern grey kangaroos slowly hopping across pastureland. (Photo: Dave Watts.)

from the elastic recoil of their musculoskeletal system. When hopping, the tail and leg tendons store energy when the kangaroo lands and then release this energy during the following leap. Some elastic energy storage and recovery is accommodated by the bending of long bones. The tail is not only important as an energy recoil structure but also important as a sensor for body position. What is more, it counteracts the pitching rotation of the trunk, arms and head associated with a hopping gait. He found that when red kangaroos hop at speeds between 15 and 35 km h^{-1} the stride frequency is constant, but the stride length is varied appropriately. However, with speeds above 35 km h^{-1}, stride frequency as well as stride length is increased. While the energy needed to start hopping is high, the ongoing energy costs to maintain speeds between 15 and 35 km h^{-1} are about one-third of that used by a quadruped of equivalent body mass. Professor Dawson also found that kangaroos have as high a metabolic and aerobic capacity as comparable eutherians. He suggests that the high metabolic capacity of red kangaroos is a major factor in their energetically efficient bipedal hopping between 15–35 km h^{-1}.

Other body structures are affected by bipedal hopping. Macropodids inhale as they leave the ground and exhale on landing. This is facilitated by a number of visceral adaptations. In most mammals the stomach is tightly bound to the diaphragm, but in macropodids and potoroines, after the oesophagus traverses the central tendinous part of the diaphragm it extends a long way into the abdomen before connecting to the stomach. In addition, the stomach and intestines have long mesenteric attachments allowing them to move readily. When a kangaroo propels itself upward at the beginning of a hop, inertia pushes the viscera backwards within the abdomen, simultaneously pulling the diaphragm backwards. This helps to draw air into the lungs. When the animal lands, the viscera are pushed up against the diaphragm, forcing air out of the lungs. This piston-like action facilitates inspiration and expiration. During hopping, the respiratory rhythm is phase locked with the rate of hopping in a ratio of 1:1. Heart rate is also linked to locomotor activity but in a less rigid manner. Here cardiac frequency rises at a faster but steady rate as locomotor speed increases.

Arboreal kangaroos

Tree kangaroos only weigh about 75 per cent of their similar-sized terrestrial counterparts, and much of their leg and lower back musculature has been redistributed to the upper body.

Essential to a tree kangaroo's arboreal lifestyle is great upper body strength, which allows it to pull its body up when climbing. This is reflected in a significant increase in its forelimb muscle mass to about 25 per cent of the total body mass (TBM). There is also a reduction in the mass of the lower trunk musculature to about 33 per cent of TBM, and a shift of muscle mass to the upper trunk to assist with the demands of pulling the body upwards. These changes have shifted the centre of mass forwards compared to their terrestrial counterparts. Although their long tail is not prehensile, it is a significant balancing agent.

Tree kangaroos' arboreal lifestyle demands extreme body coordination and balance. They have evolved long, robust forelimbs to reach forward and overhead as well as to be able to pull their body upward using a backwardly directed power stroke. To facilitate this, their shoulder joint has interconnected shoulder tendons forming a 'rotator cuff' that holds the humeral head firmly in the scapula's glenoid (shoulder socket), thus ensuring the joint's stability. Their highly flexible wrist joints enable the wrist to rotate inwardly when grasping a tree trunk or branch. Long, robust, curved claws plus palms with large pronounced pads covered by uniform toughened tubercles give them a sure grip of the substrate.

Tree kangaroos also have a number of subtle hindlimb adaptations that facilitate their arboreal lifestyle. The tibia and fibula are relatively short and are separated, allowing them to have a high degree of rotation. The tibial articulation with the talus (the upper bone of the ankle joint), the principal weight-bearing anklebone, is smooth, allowing some lateral movement. Their largest anklebone, the calcaneus, has a smooth broad articulation with the talus above and central tarsal bone below. These adaptations, together with smoothing of articulations in the lower ankle joint, allow tree kangaroos to be able to rotate their foot inwards to help grip the tree that they are climbing.

Tree kangaroos' feet are short and broad with long, strong, curved claws. Their footpads are supple and have a granulated form that improves their contact surface area on the substrate. Tree kangaroos are able to move their limbs either independently or synchronously when climbing.

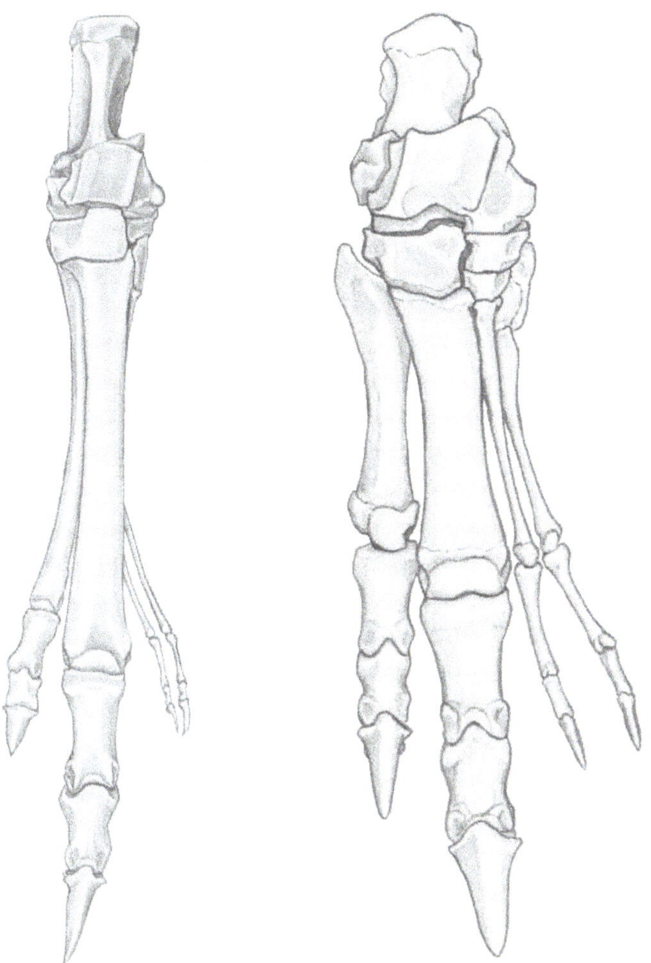

Figure 2.9: Comparative foot skeletal anatomy of the western grey kangaroo (left) and Bennett's tree kangaroo. The arboreal tree kangaroo has a foot that is much more robust and is relatively shorter than that of the terrestrial western grey kangaroo. (Drawing: Natalie Warburton.)

Adaptations for a herbivorous diet

Foraging macropods are always highly vigilant. Protruding ears are elongate in species dependent on auditory cues and short in species more reliant on vision and olfaction. Eyes placed laterally at the junction of the snout and cranial vault allow good vision forward and to the side but limit binocular vision. The elongate snout places the nostrils as well as the incisor teeth well forward. The extended position and narrow incisor arcade allows good access to the base of grasses as well as to plant extremities. Immediately behind the incisors is a diastema, an elongate region of the jaw devoid of teeth. Behind this are the cutting and grinding teeth, the premolars and molars respectively. The tongue has little involvement in food acquisition but is involved in manipulating food around the mouth during chewing and when swallowing.

Macropodiformes consume a wide variety of plant species, ranging from grasses (monocotyledons) to flowering plants (dicotyledons), ferns and fungi. Even so, of the vast array of plants available in their environment, each kangaroo species is highly selective of what it prefers to eat and what it avoids.

The small kangaroo species – those weighing 5 kg or less – are the most selective, because they need substantial amounts of high-quality food to satisfy their high metabolic rate. Consequently, small species select for more nutritious food items such as fruits, seeds, flowers, growing shoots and fungal fruiting bodies: dietary items that are often spread over relatively wide areas and are generally patchily distributed. Browsers feed primarily on plant material that has a lower fibre content than that selected by grazers. The larger kangaroo species are the least selective: their greatest need is to harvest enough food to supply their daily needs.

The structural strength of plants is dependent on the arrangement of their polysaccharide–polyphenolic building blocks of cellulose, hemicellulose, pectin and lignin. Young plants have mostly cellulose and hemicellulose cell walls confining their cellular contents, but as plants mature, more and more indigestible lignin is laid down, confining the cellular contents. Generally as plant parts age, especially in pastures, their nutritional value declines. Browsing kangaroos harvest leaves, twigs, shoots, flowers and fruiting bodies mostly from dicotyledonous plants. Dicotyledons have their major supporting structures, the vascular bundles, arrayed irregularly. Browsing kangaroos use their incisor and premolar teeth to apply high initial shear forces to break larger plant material into smaller units. Then fine shearing in the molar arcade reduces the food further. However, grazing kangaroos primarily harvest grasses. Grasses

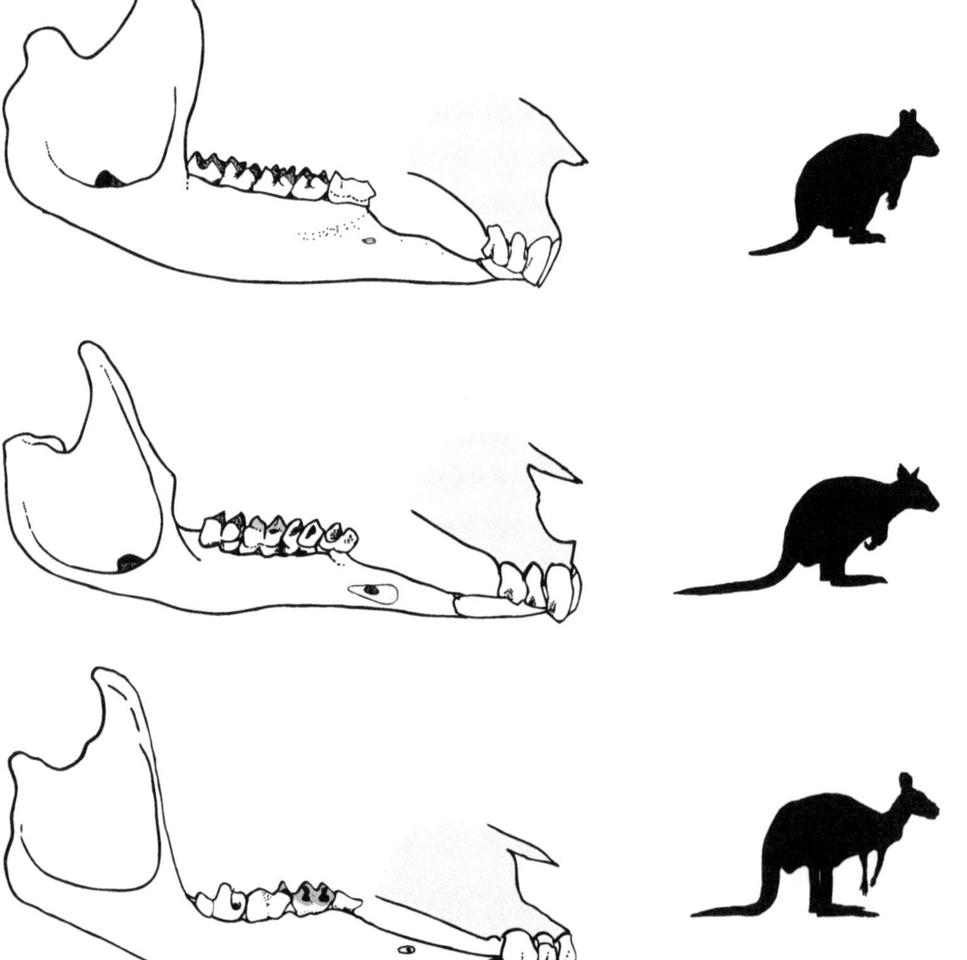

Figure 2.10: Original drawings following the concept by Sanson in *Kangaroos, Wallabies and Rat-Kangaroos* (Grigg *et al*., 1989). Top image is of a typical browser, the quokka, showing the straight lower molar arcade ending with a large sectorial premolar. Their lower incisors move medially to upper incisors. The middle image is of a grazer-browser, the tammar wallaby. This has a straight molar row and the small premolar is shed. Lower incisor occlusion is along the medial aspect of the upper incisors. Lower image is of a typical grazer, a western grey kangaroo. Here two molars are 'in wear' and are positioned on an elevated section of the mandible. Lower incisors bite against large upper incisors. (Drawing: Perdita Phillips.)

differ significantly from dicotyledonous plants in having proportionately a higher fibrous component that is distributed uniformly and is extremely resistant to physical breakdown.

Because macropods do not regurgitate digesta for secondary mastication, they must masticate their food well to maximise the availability of the plant cell contents for digestion, and also so that the cell walls are broken down into small particle sizes to allow the most efficient microbial digestion possible.

Some kangaroo species, for example red and grey kangaroos, are primarily grazing animals, while others, such as tree kangaroos and the quokka, are primarily browsing animals. Another suite of species, such as the pademelons, is intermediate, in that they both graze and browse extensively. So what are the morphological attributes that facilitate these differing feeding methods? In fact there are a number of variables involved. These are: the shape and size of the upper incisor arcade, as well as how far the lower incisors protrude towards the upper incisors and whether or not they contact these; the shape, size and alignment of the premolars; whether the cheek tooth row is flat or inclined (the angle of inclination determines whether 4 or 8 molars on each side are in contact with one another); and the extent of contact between the molar occlusal surfaces.

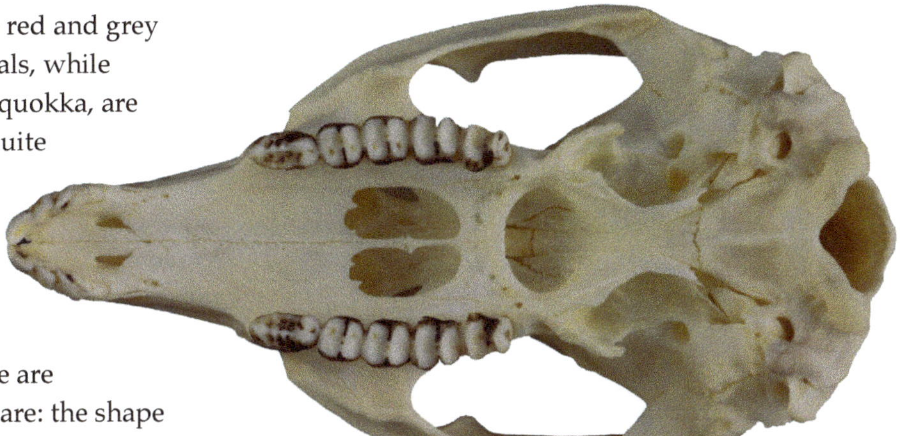

Tooth categories

Incisors

All kangaroos have three pairs of upper incisors and a single pair of lower incisors. In the upper jaw, browsers have a pair of large central incisors that are each flanked by two smaller incisors. The lower incisor does not protrude as far as the upper incisors, and bites onto a tough horny tissue pad that lies immediately behind and adjacent to the upper incisors. Browsers shear some plant material between their upper and lower incisors.

Contrasting with this is the arrangement seen in the upper jaw of the grazers, especially the large grazing

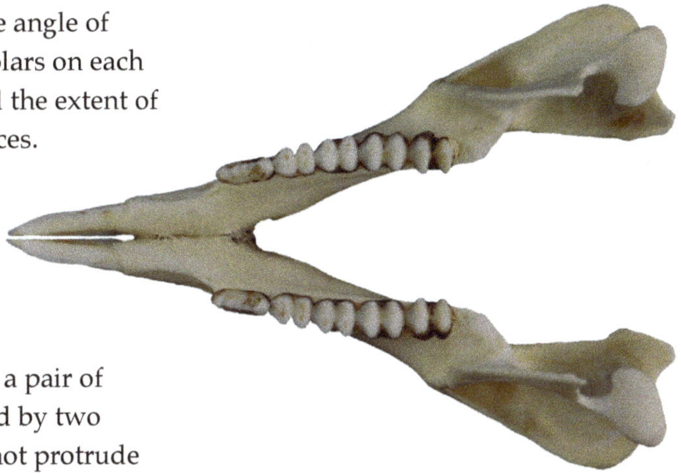

Figure 2.11: This quokka dentition is typical of browsing macropods. Its lower procumbent incisors shear browse by moving on the inner side of the upper three pairs of incisors. Quokkas also have robust sectorial premolars that scissor past each other when shearing browse. (Photo: Joe Hong.)

species. Here the corner incisor is the broadest, the laterals are quite small and the centrals are a little larger than the laterals and more protuberant. The lower incisors extend forward, forming an elongate occlusal surface that acts along the entire upper incisor arcade facilitating close cropping of grasses.

Canines

Where present, canine teeth are small, short and have little to no role in food acquisition and processing. They are present in most genera of kangaroo except *Macropus*, *Petrogale*, *Wallabia*, *Thylogale* and *Setonix*.

Premolars

Browsing kangaroos use their premolar teeth to apply high initial shear forces to break plant material into small units. Adult browsers retain a single large permanent premolar tooth immediately in front of each molar tooth row. Relative to the size of their skull, these premolars are large and elongate. In most browsing species they are aligned along the length of the dental arcade and have a chisel-shaped occlusal surface. Because the upper dental arcade is wider than the lower one, the inner surface of the upper premolars slides closely over the outer surface of the lower premolars. This enables an effective shearing action on twigs and stems. In contrast, grazers do not retain their premolar teeth. Their premolars are small and are usually shed before adulthood. This allows the anterior migration of the molar teeth as the animal ages.

Molars

Browsers have a flat cheek tooth row, with all four molars in each quadrant in wear for most of the adult life of the animal. The individual molars have high transverse lophs. In occlusion the opposing lophs slide into the deep transverse troughs as far as the small longitudinal ridges that link the anterior to posterior lophs. Consequently, during occlusion the

Figure 2.12: The Tammar wallaby dentition is typical of a grazer–browser. It uses its lower incisors to bite directly on the medial aspect of the upper incisors. This creates an elongate cutting edge for it to bite off grass. The small premolars fall out early in life. (Photo: Joe Hong.)

surface area of contact is large, resulting in crushing of the plant material. In grazers, the lower cheek tooth row has a vertical convex arc that allows maximal occlusion pressures and hence efficient processing of their graze. In the larger grazers, such as the red and grey kangaroos, the vertical convex arc (the curve of Spee) is extreme, particularly in older animals, resulting in only two lower molars on each side occluding at any one time with the upper molars. Each molar has transverse lophs with relatively high longitudinal linking ridges. Consequently, during occlusion the molars can only slide a short distance into each other, acting more like cutting devices than grinders. Grinding action is from the front to back. In grazers the anterior molar progression over an animal's life ensures that it has molar teeth capable of processing its food. Browsing-adapted teeth have greater dietary latitude than grazing-adapted teeth. Hence browsers can adapt to grazing more readily and efficiently than grazers to browsing.

Since the upper dental arcade is significantly wider than the lower dental arcade, food acquisition (incisors, premolars) and subsequent chewing (molars) is completed on one side and is then transferred to the other side. Grazing tammar wallabies bite more often than when they are browsing. Their efficiency in comminuting ingesta is better when grazing than when browsing.

Form and function of the alimentary tract

Kangaroos' possession of a large, tubular stomach is the critically important and unique structure that has allowed the Macropodiformes, except for the musky rat kangaroo, to exploit their environment successfully using plants as food. Food is retained in their stomach for about two-thirds of its total transit time through the alimentary tract. Professor Ian Hume, a pioneer in the study of stomach and intestinal form and function in marsupials, found a major function of the kangaroo

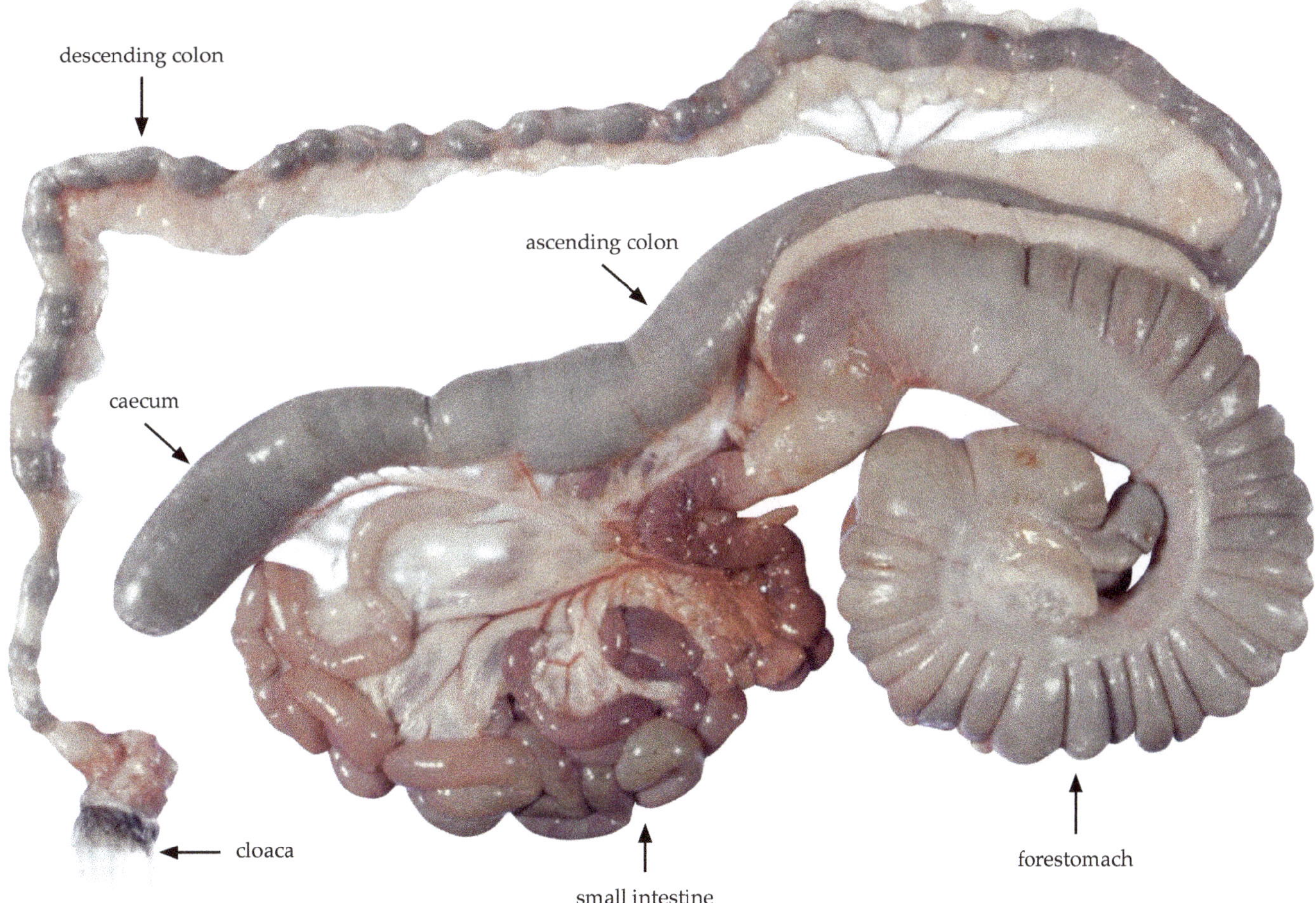

Figure 2.13: The digestive tract of a tammar wallaby. This consists of a voluminous tubular stomach and a long small intestine leading into the large intestine. The latter consists of a simple but large caecum for secondary fermentation and the colon. The terminal colon contains many faecal pellets and terminates in the cloaca. (Photo: Geoff Griffiths.)

stomach is to separate the liquid and solid phases of the diet. In the red-necked pademelon, tammar wallaby and eastern grey kangaroo, the fluid phase of the digesta passes through the alimentary tract in roughly 20 hours, whereas it takes up to 45 hours for most of the solids to pass through.

The stomach of the Macropodiformes is a voluminous, elongated sigmoidal tube, with its initial part being the largest (90-plus per cent), a non-glandular forestomach where microbial fermentation occurs. The terminal part, the glandular hindstomach, is where true digestion commences.

The forestomach consists of a sacciform and a tubiform compartment. The blind ending sacciform compartment is the region forward of the oesophageal entry into the stomach, and the tubiform compartment leads towards the hindstomach. The volume of the tubiform compartment of the complex ranges from 52 per cent in the red-necked pademelon to 72 per cent in the eastern grey kangaroo and 76 per cent in the Bennett's tree kangaroo.

Among the different species, the luminal lining of the forestomach consists of varying proportions of stratified squamous epithelium and cardiac glandular region. Only the cardiac gland mucosa is secretory and produces mucins, hence the pH of the region is mostly due to microbial by-products. Effectively the forestomach is simply where food is held for prolonged periods to facilitate anaerobic microbial fermentation mostly by bacteria (10^{10} g^{-1} of digesta) and to a lesser extent protozoans (10^4 g^{-1} of digesta) and fungi. As food passes along the forestomach it is mixed constantly so that the microbes are exposed continually to undigested plant material, and thus improve digestibility. These microbes produce a variety of enzymes, the most important being cellulase (not produced by mammals), that break the plant's cell wall carbohydrates – cellulose, hemicellulose and pectin – down into smaller structural units. These are incorporated into the microbes, allowing their growth and reproduction. As the microbes break down the plant cell walls they release a variety of short-chain

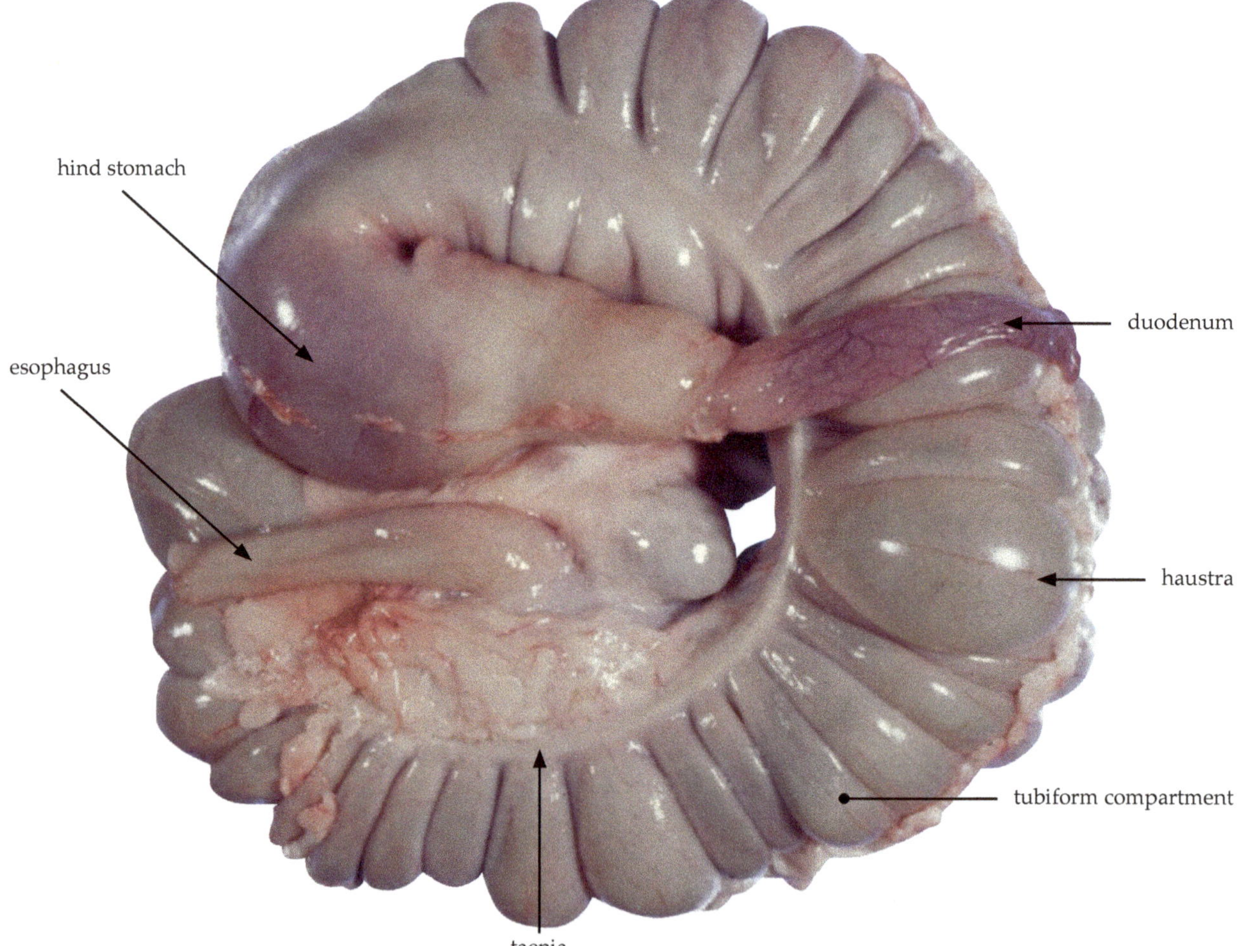

Figure 2.14: The right side of a tammar wallaby stomach. The numerous haustra (sacculations) are essential for mixing the dietary components. The hindstomach is the 'true' stomach, where the microbes are killed and their digestion commences. (Photo: Geoff Griffiths.)

fatty acids that pass through the kangaroo's stomach lining and are absorbed directly into the blood, traversing within the blood vessels of the stomach wall. Kangaroo microbes use acetate as a substrate in their growth and multiplication. Hence kangaroos on a 'rough' diet such as dry fibrous grass have a 10–15 per cent advantage in energy availability and thus are more efficient than ruminants. However, on improved pastures they have less of an advantage. Kangaroo microbes produce little methane compared to ruminant microbes, which lose up to 15 per cent of their potential energy this way. When digesta passes into the highly acidic environment of the hindstomach the microbes are killed and proteolytic enzymes immediately commence the breakdown of the microbial protein.

Once digesta passes into the small intestine, the alkaline environment facilitates the breakdown and absorption of microbial products and of any available plant; fat, protein and carbohydrate. By this stage there is little digestible carbohydrate left, it having been dealt with in the stomach. The large intestine has a simple, but quite voluminous, caecum and long colon, both of which are sites for secondary microbial fermentation. Although the fermentation produces significant levels of short-chain fatty acids, it is probable that this region is more important for water reabsorption. Grazers have longer large intestines than do browsers. The colon of arid-adapted animals is proportionately longer than their more mesic-adapted counterparts.

How a pouch young obtains its strictly anaerobic microbial flora in the first place, especially the protozoans, has not been determined conclusively. It is probable that, following the mother's grooming and cleaning of the pouch, the pouch young incidentally ingests microbes from the mother's saliva. In the routine husbandry of orphan pouch young tree kangaroos (notably from road-kill females) it has been found essential to give them an oral inoculum of microbes obtained from the mother's stomach, so that the joey can cope with a herbaceous diet.

The role of differing dietary choices

Strict grazers are generally the larger species (more than 30 kg), such as the red kangaroo, eastern grey kangaroo and western grey kangaroo, as well as the common and antilopine wallaroos. These animals need less food per unit body weight because of their body's lower metabolic rate than is the case with their smaller counterparts. This allows the larger males, in particular, to cope well on low-quality food. Consequently, on the same pasture the males are less selective than the smaller females. In arid areas female red kangaroos will be selecting high-quality plant species as well as higher quality plant parts. Selection for 'soft' grasses – those that have a low fibre content – and avoidance of highly fibrous plants, such as dried-off grasses and the highly sclerophyllous spinifex grasses, is usual.

As seasonal pasture quality drops through increasing lignification and decreasing cellular contents, all large macropods become less selective. Later in the season, when favoured grass species are either scarce or absent, animals have to eat less favoured species. Hence one can observe animals switching from selecting low-fibre, high-protein grasses to selecting higher-fibre and lower-protein content grass species. The larger macropods prefer grasses, and will only take perennial plants such as bluebush when more favoured plants become unavailable on overgrazed habitats, and notably during severe droughts. Over large areas of Australia spinifex is abundant; however, it is highly lignified, and as the plants age the lignin increasingly surrounds the cellulose, thereby decreasing its nutritive value, not to mention making it a lot tougher to chew. Spinifex has a nitrogen content varying between 0.06 and 0.5 per cent. Only the euro can survive for extended periods on such a poor diet. Fibre digestion is less efficient in smaller macropods than it is in larger species.

Browsing is practised to some extent by most kangaroos. Species such as the red-necked pademelon, quokka and swamp wallaby are primarily browsers, even though they all consume some grasses in their diet. Where possible, they browse softer plant parts of high nutritional value but low in fibre. The amount of browse eaten varies, from 70 to 90 per cent in quokkas, dependent on season; to about 20 per cent in black-striped wallabies that only take a significant proportion of browse during severe drought. Both grazer–browsers and browser–grazers spend less time feeding per gram of dry matter intake when browsing compared to when they are grazing. There is a suggestion that the chewing frequency is faster when browsing than it is when grazing.

The smaller kangaroo species must eat food that is of high nutritional value to satisfy their high metabolic rates. The smallest member, the musky rat kangaroo, is the only kangaroo species to have a simple stomach and not be reliant on the processes of fermentation and hence microbes for the digestion of its dietary items. These animals are mostly frugivorous, and supplement their diet with nutritious seeds and invertebrates. The other small kangaroo species are 'concentrate selectors' that have a diet low in fibre. All of these have the typical voluminous forestomach for dietary fermentation. They are a disparate group of species ranging from the quokka, swamp wallaby, hare wallabies and rock wallabies, to potoroos and bettongs. All are opportunistic feeders that eat varying amounts of fungi, fruits, tubers, bulbs, roots, flowers, leaves, gum exudates, seeds and invertebrates. The dietary mix varies greatly seasonally. Many bettongs and potoroos are highly dependent on truffles (mycorrhizal fruiting bodies) that at times may constitute up to 90 per cent of their diet. When they locate a truffle they eat its soft inner contents and leave the outer tough parts behind. Truffle lipid levels are high and provide a source of readily available energy. Truffle structural carbohydrates are high in concentration and supply substrates for microbial fermentation. Bettongs frequently chew dietary items such as seeds, swallow the soft nutritious contents and spit out the macerated extraneous roughage. When eating arthropods they suck out the soft contents and discard the exoskeleton.

Tree kangaroos spend their time browsing in forest trees and vines. Of necessity they are highly selective folivores avoiding mature forest leaves that are low in nutrients and are frequently packed with toxic secondary metabolites (tannins, phenolics and terpenes). The absolute levels of the toxins vary with the plant species, the stage of growth and the seasonal conditions. High concentrations of secondary metabolites can be directly toxic to the animals or disrupt their microbial fermentation. Consequently tree kangaroos avoid eucalypts. When foraging they select young growing shoots because these have high crude protein and phosphorus levels coupled with less cell wall structural carbohydrates. They also seek out flowers and fruits of many plant species when these are available.

The fibrous nature of the diets of all kangaroos means that as they get older their molars become worn, resulting in poor mastication of their food and often starvation. Remarkably, one species of kangaroo, the nabarlek, a small rock wallaby, has continual molar tooth replacement over its lifetime. No one knows how many molar teeth are used in its lifetime over and above the 4–5 per dental quadrant found in all other Macropodiformes. However, it is a remarkable variation that may confer adaptive advantages on these small rock wallabies.

Professor Gordon Sanson, who studied kangaroo dentition extensively, suggested that this may have arisen in response to nabarleks eating large amounts of the extremely highly siliceous fern *Marsilea crenata* (18 per cent dry matter). Grasses in the area also have seasonally high levels of silica. However, the nabarlek is sympatric with the short-eared rock wallabies and the monjon over much of their distribution, and neither of these two species has this unique adaptation. This unique adaptation is difficult to explain. One possibility is that there has been a random genetic change affecting one to two of the genes responsible for tooth development, which has resulted in a developmental switch causing a different phenotypic outcome to that of all other members of the Macropodiformes.

Feeding ecology

Kangaroos have a field metabolic rate that is about 70 per cent that of comparably sized herbivorous eutherians. Consequently, their daily food intake is less than that of their similar sized eutherian counterparts. Even so, a major component of the daily activity budget of all wild kangaroos is foraging to meet their daily food requirements. Except for the diurnal musky rat kangaroo, most kangaroos feed during the night and, in some cases, may extend this into the early morning and late afternoon.

Foraging behaviour is influenced by factors such as the body size of the animal, habitat type and climatic zone. Body size determines metabolic rate, which in turn determines the quality and quantity of the food needed to maintain the animal. The smallest kangaroo, the musky rat kangaroo, weighs less than 1 kg as an adult, and has a great physiological demand for a diet high in available protein, fats and carbohydrates. Since these kangaroos do not rely on the fermentation of their diet they are primarily frugivorous and secondarily insectivorous and mycophagous. All other kangaroos have diets containing a high proportion of plant structural elements, and rely on a voluminous stomach where microbial fermentation occurs, converting plant polysaccharides into nutrients available to themselves.

Most potoroines have an adult body weight of less than 2.5 kg, and consequently have a relatively high metabolic rate. Of necessity they must forage for high-quality food, such as the underground storage elements of plants (roots, tubers and truffles), herbs, fruits, seeds and invertebrates. For most potoroines, much of their foraging behaviour is governed by truffle availability. They move rapidly and, if necessary, over long distances in their quest for patches of nutritious truffles. They find them by their olfactory signature. When they find an area where truffles are present, bettongs and potoroos move slowly and in short steps to increase their probability of locating one. When they find a truffle they use their forelimbs to dig down to it.

The next size category consists of kangaroos having adult body weights between 2.5 and 12 kg. This is a mixed group, with the smaller species being mostly browsers and the larger species tending towards being mostly grazers. Most of these species exhibit some form of dietary selection. Selection ranges from consuming only dicotyledon leaves through to only grass leaves, through to being non-discriminate feeders of all parts of grasses.

Most macropods having a body weight of greater than 12 kg exhibit some dietary selectivity, depending upon seasonal conditions and their preferred habitat. Metabolically these animals need fewer nutrients per gram body weight to maintain them than do their smaller family members. Consequently they are able to eat large volumes of lower-quality plant material. Even so, smaller individuals as well as lactating females need a higher quality diet, and hence are more selective of their diet.

The larger kangaroo species have a routine daily activity pattern that allows them to take advantage of favourable environmental conditions and simultaneously minimise the effects of adverse ones. A typical day's activities include moving to and from their main grazing areas, foraging movements, feeding and social interactions. In the heat of summer, red and grey kangaroos forage for only about 40 per cent of their daily time budget, but in the coldest periods of winter this may rise to 80 per cent. The limited activity in summer is an attempt to avoid heat during the day and reduce physical activity during the night to lessen energetic movements that elevate the animal's core body temperature. At this time of year their aim is to start the day as cool as possible to minimise the inevitable rise in core body temperature for as long as possible. In winter, foraging takes up the majority of the daily activity budget, mostly to obtain enough food to cover energetic demands such as shivering to maintain body heat.

Another important variable influencing their daily activity pattern is the changing plant phenology, with varying stages of plant growth and death resulting in different amounts of available plant water, nutrients and structural carbohydrates. The quality and quantity of suitable food determines foraging time. In the field, the availability of particular plant species will modify the actual dietary intake of an animal. While an individual kangaroo species may prefer a suite of plant species, if these are low in abundance it may not be in the animal's best interests to spend extra time seeking out the preferred plants. In other words, the energetic cost involved in searching for the preferred plants may be more than the nutritional and energetic return from consuming them. Consequently it may make more sense to forage on less desirable plant species.

We have seen that dentition is a critical factor in the determination of a macropod's diet. Species such as tammar and agile wallabies retain relatively large premolars that allow them to browse; however, they are more efficient grazers. When browsing, tammar wallabies spend proportionately more of their day chewing than they do when grazing. There is a trade-off between the higher qualities of the browse against a poorer oral processing of these food items. Because of their limited dexterity, they frequently drop the leaves that they are holding to eat. Dropped leaves are left where they lie. Overall, tammar wallabies spend more time eating smaller volumes of food than do their much larger relatives, the red kangaroos.

With most macropods, feeding activity peaks in the early evening and again just before dawn. This is a result of an interaction between the distribution of the food resources and the day-night variation in the animal's neural feeding drive. Female red kangaroos may graze for up to 8.5 hours per day in summer; however, 6 hours is closer to the norm. Since their metabolic demands are greater than those of the large males, the females are more selective for high-quality green grass. Green grass shoots are much superior in their nutritional quality and digestibility than dry coarse feed. Red kangaroos prefer green, highly nutritious succulent grass shoots, while grey kangaroos are better adapted to a diet of coarse grasses of lower nutritional value. Digestive efficiency among large kangaroo species varies, with that of western grey kangaroos being better than that of the eastern grey kangaroos, and they in turn have a better digestive efficiency than the red kangaroos. Depending on distance, predation risk and other factors, such as thermal conditions, all large macropods will move moderate distances to better-quality pasture wherever and whenever possible.

Adaptations to maintain body fluids and temperature

Water balance

Kangaroos have adapted to the challenge of water balance in Australia's arid deserts through to wet rainforests. However, there is nothing remarkable about the overall structure of their urinary system. Even so, water can be a significant factor in the welfare of individuals of different species when water availability is either over abundant or limited. Water excess is rarely a problem. However, after long dry periods followed by rain the water content of new plant growth can rise to 90 per cent of fresh mass. Under these conditions, to meet their nutritional needs animals ingest large amounts of vegetation that is basically 'waterlogged'. Tammar wallabies, agile wallabies and quokkas can lose body condition, and on occasion even die under these circumstances.

In the eucalypt forests of northern Australia, where water availability is usually not constraining, rufous rat kangaroos have a uniform water intake throughout the year, primarily from dietary sources. However, where rainfall is highly seasonal, species such as tammar wallabies, quokkas and brush-tailed bettongs all have intakes that follow the availability of both dietary and free water. Their water intake rises during the months of good rainfall, and when rain is scarce or absent it drops accordingly.

Some species of kangaroo, such as the red kangaroo, the much smaller common wallaroo and the even smaller rufous bettong, are well adapted to live in Australia's challenging hot arid zones, but

Figure 2.15: Red kangaroos drinking from edge of the Murray River. When water is readily available these large kangaroos will drink regularly. Otherwise they adjust their water and heat conservation to accommodate water availability. (Photo: Jiri Lochman.)

Figure 2.16: Landscape west of Cue in Western Australia. Caves and overhangs of these 'breakaways' are typical euro 'lie-ups' during the heat of the day. (Photo: Marie Lochman.)

less so to the moist and often more humid coastal areas. All three species have or had large geographic distributions over the arid–semi-arid zones.

Arid zone macropodids have several physiological and behavioural responses that help them cope with the rigours of living in an environment that can be extremely hot and dry. Red kangaroos rest up during the heat of the day, generally in the shade of a tree or occasionally in the shadow of a rocky outcrop. When resting, they cool off using evaporative heat loss by panting and licking their forearms. During periods of extreme daytime heat, water loss facilitating this evaporative cooling is high. Professor Terry Dawson, working at Fowler's Gap in western New South Wales, reported that water consumption in adult red kangaroos increases from the cooler periods (July average minimum 5°C to maximum 15°C) to the hotter periods of the year (January average minimum 18°C to maximum 33°C) by about 55 mL per kg body weight per day. When conditions are very hot, red kangaroos often adopt a sitting position hunched over, with their tail placed forward between their legs. This position allows air circulation around their body while also minimising their body surface area exposed to solar radiation. Another coping strategy they adopt is to drop their core body temperature by 2–4°C in the hours preceding the heat of the day. This lessens the total period spent each day with an elevated core body temperature. Throughout hot periods they have low daily activity patterns that extend into the evenings when they remain less active than they are in the cooler months. This results in less heat production from exercise as well as keeping the overall body temperature as low as possible at all times. Red kangaroos can tolerate a greater body heat accumulation over the hottest part of the day than the grey kangaroos.

Other kangaroo species use a variety of different strategies to cope with hot weather. For instance, during the hottest part of the day euros often crawl into the caves found at the base of rocky outcrops, where the temperature may be up to 10°C cooler than the outside ambient temperature. Many rock wallaby species do likewise. The yellow-footed rock wallaby appears to be an exception, in that over the heat of the day it prefers to lie up in the shade of shrubs rather than shelter within a rock pile.

The percentage of water within the diet influences

Figure 2.17: Another typical view of where euros may 'lie up' during the heat of the day in the Murchison region of Western Australia. (Photo: Jiri Lochman.)

whether or not a kangaroo needs to drink. When their dietary water content falls below 55–65 per cent, kangaroos require free water. However, the need to drink varies with different species; for example, eastern grey kangaroos need to drink regularly. By the middle of most hot summers eastern grey kangaroos drink considerably more than red kangaroos, and they drink more frequently. Yellow-footed rock wallabies drink each night, while euros drink several times each week. Euros drink about 12 per cent of their body mass per week, and this is possibly a ballpark figure for all of the larger arid-adapted macropods.

During extreme water deprivation, euros preferentially maintain water levels in their gut, thus preserving their ability to digest food efficiently. They only lose 20–30 per cent of their total water loss from the gut, while under the same conditions red kangaroos may lose 50–60 per cent from the gut. Red kangaroos have a lower microbial fermentation efficiency under these circumstances.

Arid-adapted macropods have the possibility of reducing their faecal water loss when water is scarce. For instance, in euro populations the western more arid-adapted form has a colon that is 30 per cent longer than is found in the eastern form. The increase in length allows more water to be extracted from the faeces and retained in the body. Consequently, western euros can produce faeces that are up to 10 per cent drier than those of their eastern relatives.

Analyses of the distribution of two closely related macropods, tammar and parma wallabies, following their introduction to Kawau Island, New Zealand, about 140 years ago, show that the two species occupy different parts of the island. This is believed to be because tammars have a greater renal water-conserving capacity (longer Loops of Henle) and better colonic water resorption (longer colon), resulting in their needing less free water. Consequently the tammars colonised the drier areas of the island.

Another mechanism available to kangaroos to minimise water loss is to recycle urea, a waste product of protein breakdown. Normally urea is excreted through the kidneys in a process that is dependent on water transportation. In red kangaroos, under cool conditions and when free water is readily available,

Figure 2.18: A mummified euro in its 'lie-up' in a cave. (Photo: Jiri Lochman.)

the urine osmolality is about 1.5 mOsmol kg^{-1}. Under such conditions, a lot of water is used to remove the urinary excretory compounds. However, in hot conditions coupled with limited available water, urea is reabsorbed from the kidneys into the bloodstream and less water is used, resulting in significant water savings. Under severe conditions, urine concentrations of up to 4.1 mOsmol kg^{-1} have been recorded for red kangaroos. This is much higher than that achieved by eastern grey kangaroos, where urinary concentrations may reach 2.8 mOsmol kg^{-1}. Reabsorbed urea is secreted from the blood into the saliva, swallowed and used in the stomach as a nitrogen source for the microbes. Hence the stomach's microbial ecology is improved and there is better digestion of the available poor-quality forage.

Both the red kangaroo and euro have the remarkable ability of being able to maintain their blood plasma levels close to normal as their overall body dehydrates severely. Thus their blood does not thicken and place severe stress on their cardiovascular system.

Within the five species of small macropod in the arid Pilbara region of Western Australia there are three significantly different solutions to retaining body water. On Barrow Island (average rainfall 307 mm), where summer temperatures commonly exceed 35°C and often rise above 45°C, and daily relative humidity ranges between 50 and 65 per cent, the black-footed rock wallaby and the burrowing bettong use behavioural strategies to avoid the extremes of daily thermal stress. They remain in cool and humid caves or deep burrows during the day, resulting in their maintaining a constant body temperature, having a low evaporative water loss, having a high daily rate of water turnover and producing relatively dilute urine. In contrast, the euro and spectacled hare wallaby shelter in shady places to lessen the thermally stressful ambient conditions common to Barrow Island. These animals have a low daily rate of water turnover and produce highly concentrated urine. This is facilitated by their finely tuned physiological control of water homeostasis via antidiuretic hormones regulating water reabsorption in the kidneys. Interestingly, Rothschild's rock wallaby, another Pilbara species, uses both behavioural and physiological means to limit water loss. It avoids extreme thermal stresses by remaining in caves and caverns within rock piles where the relative humidity is high and the temperature is up to 10°C less than the ambient air temperatures in the surrounding area. Also, they limit their water loss at the kidney level by reducing blood flow through and having a low filtration of the blood at the glomerulus. Thus in times of heat and water stress they do not concentrate their urine beyond normal level; they simply void less.

Body temperature

Adult and sub-adult macropods endeavour to maintain a core body temperature of 36°C. While their temperature regulation is excellent their ability to cope with climatic extremes varies with species and geographic locality.

Extremes of heat pose a different suite of problems for macropods, many of which have been discussed earlier in this chapter. Red kangaroos, probably the best adapted of all macropods to hot arid environments, have short fur that is highly reflective of incident radiation. While about 70 per cent of incident radiation is absorbed into the fur, little is transmitted to the skin surface. Additionally, they have a well-developed vascular rete, on the inner surface of each forearm, that is a very effective thermoregulatory structure. When the ambient temperature rises to extreme levels, individuals lick their forearm continuously. The saliva evaporates, cooling the blood within the nearby underlying blood vessels, and this in turn cools the body core. Under severe conditions young animals are most at risk of dying. Note that many mechanisms used by macropods to limit water loss are also mechanisms to cope with heat stress.

Under normal seasonal conditions in southern Australia, kangaroos gradually replace their short summer coats with a longer denser coat as winter approaches. As the average daily temperatures drop, they increase their dietary intake to assist them to maintain their core body temperature. Time spent foraging usually increases. Notably in the cooler periods, wind intensifies the problems of cold and/or wet conditions. Wind disrupts the lie of the fur, lessening the fur's insulative effect, resulting in heat loss. Red kangaroos are affected more by lower temperatures than sympatric grey kangaroos, which have fur that is generally longer, denser and more water repellent, even in summer. Kangaroos also respond to cold conditions by limiting their

Figure 2.19: Red kangaroo reclining in a shady spot. (Photo: Dave Watts.)

Figure 2.20: Red-necked wallaby in snow in alpine Tasmania. These wallabies have denser and longer fur than their mainland counterparts. (Photo: Dave Watts.)

blood supply to their extremities, by shivering and by avoidance. In the latter instance, animals move into shelter such as gullies, or in the lee of a rocky outcrop to avoid cold, strong winds. In extremely cold circumstances animals do not come out to graze, but instead stay within whatever shelter they can find. Young animals are always highly susceptible to inclement cold weather. Red kangaroos in the southern limits of their distribution are particularly susceptible to very cold weather.

Adaptations for reproductive success

One of the most amazing features of all macropodiformes is their extremely short gestation period before they give birth to precocial, small young that are about one-hundredth the weight of their mother.

Immediately after birth, the newborn instinctively crawls unaided to and enters its mother's pouch, where it attaches to an available unused teat. From that moment it begins the long journey towards independent living. Quite distinct from eutherians, where a substantial amount of their early growth and development occurs in the uterus prior to birth, in marsupials it is within the pouch where most growth and development occurs. As a result, in the Macropodiformes the maternal investment is prolonged; for example, the time from birth to independence in the red kangaroo is about 51 weeks, in the tammar wallaby it is about 42 weeks and in the brush-tailed bettong it is about 18 weeks.

The adoption by marsupials of a strategy where much of the growth and development of their young occurs in the pouch is because of the unique anatomy of the reproductive tract, especially that of the females. The reproductive tracts in marsupials have a different arrangement of the genital and kidney ducts to the urogenital sinus and bladder to that found in the eutherians. In mammals, each ureter runs from its respective kidney to open into the distal part of the bladder. In eutherians, the embryonic development of this route is lateral to the reproductive ducts, allowing the two cervices and vaginae to fuse into single entities. However, in marsupials the distal route of the ureters passes medial to the two cervices and vaginae, preventing their fusion, resulting in the retention of these dual lateral structures. The junction of the distal cervices and the uteri is the vaginal sinus. In kangaroos, during the first parturition the vaginal sinus extends distally to join with the urogenital sinus, forming a patent birth canal referred to as a median vagina. This remains as the birth canal throughout the life of the animal. In male marsupials, the medial position of the ureters means that the testes are unimpeded during descent to the scrotum, a scenario quite different to eutherians, where the vasa efferentia are constrained to loop dorsally around the ureters before entering the common urethra.

The typical adult male macropod is an opportunistic breeder. The ancestral pattern of breeding in macropods was probably continuous, through postpartum mating, lactational quiescence and reactivation of the corpora lutea and conceptus when lactational support of the previous young ceased. Subsequent variations occurred as the species diverged in their habitat choices and exploitation of different climatic regimens occurred.

Oestrus is the period when the female's genital tract is prepared to receive and transport sperm through to the fallopian tubes as well as to receive and nourish the fertilised egg. At this stage the uterus also forms egg coats and nourishes the embryo. If a pregnancy does not occur, the uterus returns to its quiescent state, and shortly after, the next oestrus occurs and the uterus again returns to the active state. The oestrous cycle ranges from 21 days in some bettongs to 35 days in the red kangaroo to 46 days in the eastern grey kangaroo. The oestrus cycle consists of a pro-oestrus period when the Graafian follicle develops, leading to ovulation followed by the luteal phase, when the corpus luteum develops and controls the development of the uterine wall. In most macropods ovulation occurs 1 to 2 days after the onset of behavioural oestrus. It takes about 24 hours for the egg to pass down the oviduct. Oestrus in tammar wallabies occurs within 8 hours of giving birth, and lasts less than 12 hours. However, in most other macropods it occurs 1–10 days after birth.

When pregnancy occurs, the luteal phase persists until parturition occurs, and the newborn crawls up into the pouch and attaches to a teat. In the Macropodiformes, the length of pregnancy is only slightly shorter than the length of the oestrous cycle. Consequently, the female can become fertile immediately after parturition, and hence could become pregnant once again. If this occurs and there is already a newborn attached to the teat, the suckling stimulus causes the developing corpus luteum and new embryo to cease development (embryonic diapause). The suckling stimulus inhibits the growth of the corpus luteum to less than 2 mm in diameter between days 4 and 5 post oestrus. The new embryo ceases to grow at the 70–100 cell stage, and remains as a unilaminar blastocyst. The diameter of the shell membrane remains at 0.25–0.33 mm for the entire period, and mitoses are not seen. Under some circumstances the gestation can be extended for periods of up to 11 months by developmental quiescence. In the absence of lactation, females have

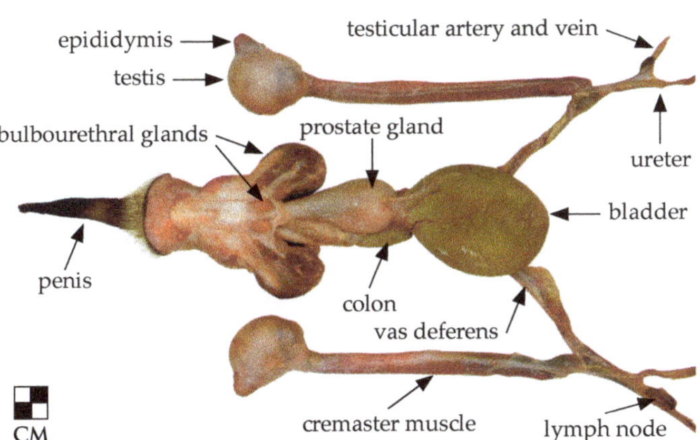

Figure 2.21: The reproductive tract of a male western grey kangaroo is similar to that of most mammals, where sperm from the testes passes along the vas deferens to the urethra, where prostatic and bulbourethral secretions complete the semen. Kangaroos have a pendulous scrotum that hangs in front of the penis. (Photo: Joe Hong.)

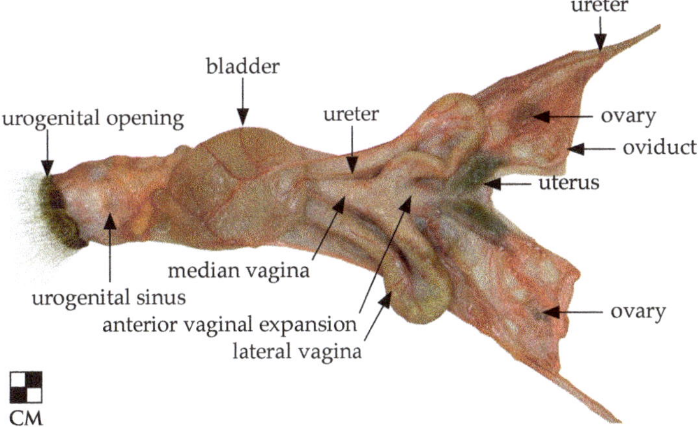

Figure 2.22: The female reproductive tract of a western grey kangaroo is characterised by having two lateral vaginae and a single median vagina. Sperm traverse the lateral vaginae and the median vagina is the birth canal. (Photo: Joe Hong.)

a brief rise in progesterone (days 6–7), and both the corpus luteum and the blastocyst develop. Most macropods have a lactational quiescence.

However, the tammar wallaby and the Tasmanian population of red-necked wallabies do not. Instead they have seasonal quiescence, which may operate for several months after the weaning of the pouch young. In these species, development resumes after the summer solstice. Tammar wallabies respond to a change in day length (long to short) with an increase in plasma progestagen and, to a lesser extent, oestradiol: a photoperiodic effect on endocrine balance.

In both seasonal and lactational quiescence, the hypothalamus inhibits the corpus luteum by the steady production of low levels of prolactin. If pouch young are lost, the absence of the suckling stimulus causes a drop in prolactin production and the reactivation of the corpus luteum. The corpus luteum begins producing progesterone. The corpus luteum increases in diameter to about 4 mm, primarily by the hypertrophy of existing cells as well as by a limited increase in cell numbers. There is also an increase in uterine endometrial secretions, and by day 8 the blastocyst resumes its development. To date, the actual mechanism for the maintenance of diapause is unknown. Note that, with reactivation of blastocyst development, organogenesis is rapid and metabolically demanding.

Professor Marilyn Renfree of the University of Melbourne has made a lifetime study of the breeding biology of the Kangaroo Island subspecies of tammar wallaby. Consequently the breeding cycle of the tammar wallaby is the best known of all kangaroo species. It is one of only two macropod species that have a strictly seasonal pattern of breeding. Most young are born from late January to March, and within a few hours of giving birth the female mates. The resulting embryo remains quiescent during lactation. The quiescent embryos are reactivated within a few days after mid December, and the young enter the pouch about 40 days later, close to 12 months

Figure 2.23: Male western grey kangaroos sparring. (Photo: Bill Belson.)

after the mating at which they were conceived. The single young is suckled in the pouch for 8–9 months, and leaves the pouch in September or October (October–November in Western Australia). Females become mature at about 9 months while they are still suckling, but males do not become mature until

Figure 2.24: Male eastern grey kangaroo sniffing a young female to determine her state of oestrous. (Photo: Jiri Lochman.)

Figure 2.25: Eastern grey kangaroos mating. (Photo: Dave Watts.)

nearly 2 years old. The rate of reproduction is high, with more than 90 per cent of all females carrying a pouch young by the end of the breeding season. In some years many pouch young are lost, particularly by yearling females. In all years, mortality is high among juveniles during their first summer, and may reach 40 per cent.

Some people believe that philopatry may direct breeding strategies. In most macropodid species it makes sense for a female to have male offspring early in her breeding life. These males will leave her and her territory to disperse elsewhere. These males are not competitors. It also makes sense for female offspring to be produced later in the female's breeding life. Then they are ready to replace her in a good environment. Until that time they will be her competitors, to some extent.

It has also been suggested that young breeding females produce more male joeys when their mothering skills are less well developed, but as they get better at raising young they produce more females later in life. This suggests that male offspring are ecologically expendable in comparison to female young.

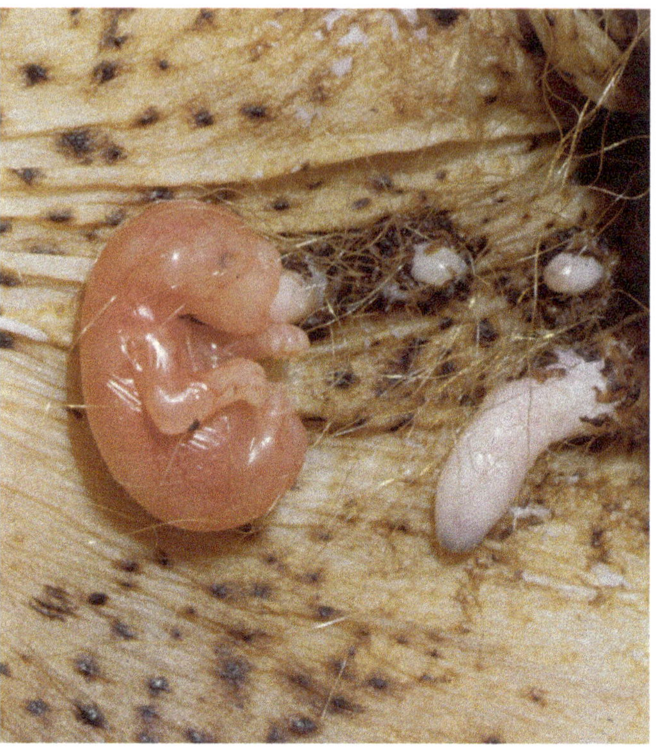

Figure 2.26: A three-day-old tammar wallaby pouch young attached to a teat. Its forelimbs are particularly well developed for it to climb from the cloaca to the pouch. There is a fourth enlarged teat utilised by young at foot. (Photo: Geoff Griffiths.)

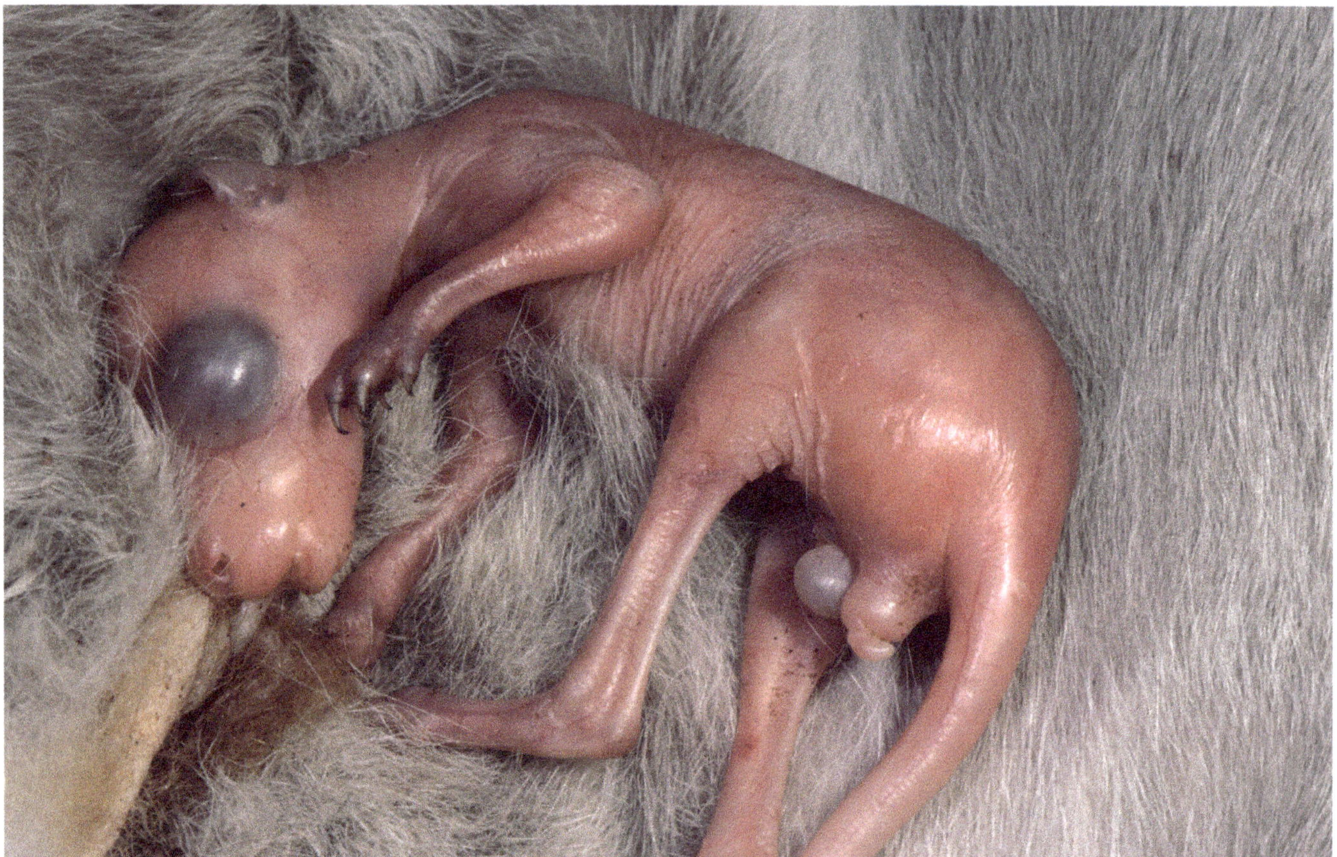

Figure 2.27: A large, hairless pouch young of a western grey kangaroo on its teat. By this stage there has been a great deal of development of limbs, tail and body systems. (Photo: Jiri Lochman.)

3. Conservation

A background to Australia's current environmental situation

Although there is no agreement on when Australia was first colonised by the ancestral Aborigines from South-east Asia, it is believed to be about 60 000 years ago. Over the following tens of thousands of years the Aborigines used patch-burning regimes to assist them with hunting as well as to refresh the country. This resulted in a mosaic of plant communities, including large areas of open savannah and woodlands that were well suited to grazing macropods. This allowed them to sustainably harvest many kangaroo species. Another major effect of their earlier occupancy of Australia is that with their hunting they may have played a role in the extinction of Australia's megafauna about 46 000 years ago.

Over the past two centuries the face of Australia has altered dramatically. Much of the continent is now threatened by serious ecosystems degradation because of past and in some cases ongoing mismanagement and neglect. This is mostly due to Europeans, whose agricultural practices evolved in environments that were fertile and robust compared with the fragile, impoverished ancient soils of Australia. When Europeans came to Australia, they brought with them a heritage that did not and does not gel well with the Australian environment. Their philosophies, which in many instances extend to modern times, encompass the exploitation of the landscape rather than accommodation and harmony.

In the early and mid twentieth century, both state and federal governments undertook strong measures to increase Australia's population and presence. Successive governments encouraged land clearing to open up large areas for agricultural enterprises. Consequently, there has been a massive increase in Australia's human population from about 3.8 million in the early 1900s to its present level of about 23 million. This in itself has had an impact on the resources available to Australia's native fauna and flora. Australia has become one of the world's most urbanised countries. Much of this urbanisation has occurred at the mouths of major waterways and where rivers meet – prime areas for large aggregations of species populations. The expansion of our cities has modified nearby land that was rich habitat. The main consequences of this have been massive habitat loss through land clearances, habitat fragmentation, climate change, soil-related changes, hydrological changes including salinity and fire, as well as the introduction of weeds, herbivores and predators. Overall, the geographic distributions of most kangaroo species have contracted, but a few of the large grazing species have expanded their range.

Within Australia there is neither the money nor the political will to correct the country's burgeoning environmental problems. Many government instrumentalities are more concerned with socioeconomic and property impacts than with the inherent value of plant and animal species at risk. Generally, governments are reluctant to make commonsense decisions in favour of the environment. All too often the overriding concern is that if they take action to remedy serious environmental issues without a socioeconomic rationale, they could alienate some of the public and potentially lose votes.

Financing and implementing environmental management of the vast areas of the public estate is beyond the realm of possibility for governments. Consequently, over recent times many individuals and in some instances corporate enterprises have assumed responsibility for long-term environmental protection of small critical areas of biological concern. Some non-profit organisations, such as the Australia Wildlife Conservancy, Bush Heritage Australia and Birds Australia, have purchased and now manage many large properties so that the animal diversity at each of these sites can be maintained 'in perpetuity'. Such enterprises are ecosystems based, and as such assist the long-term well-being of myriads of species, kangaroos included. Likewise, private and public landholders, such as rural councils and corporate bodies, have an increasingly critical role when managing land in their care to ensure the long-term well-being of the animals in their jurisdiction. Simply being aware of the presence of animals and their ongoing needs can influence informed decisions. There is an ongoing need to increase the awareness of landholders of the significant role that they can and often do play.

Initiatives such as Landcare, a national network of thousands of locally based volunteers working in community groups that collaborate with landowners, corporations and governments to repair damage to and better manage the environment, exemplify steps by government to empower people to move in this direction. With the Commonwealth Government's introduction of the 'Caring for our Country' program, significant funds became available to more than 1000 community groups and numerous regional organisations to improve environments across Australia. Many of these undertakings have certainly improved degraded habitats and advantaged small kangaroo species.

Historically, farms have played a major part in the degradation of much of Australia through such measures as inappropriate land clearances and overgrazing, and for this reason their current managers have a responsibility to be better custodians of the land. Farmers have always had an environmental subsidy of the soil components, minerals and biologicals that make up the soil profile. This is supplemented by the hydrological profile that allows such practices as cropping to continue. Farmers are totally reliant on the well-being of these for their productivity and hence success. Within Australia, about one-third of farmers are the drivers of innovation, sustainable management and profitability. The middle third endeavour to adopt and follow in the footsteps of the more successful operators, while the final third have low profitability and are mostly in survival mode. In many cases, landholders in the last category have seriously degraded properties, and are unable to deploy sufficient resources to reverse these problems.

Current and past environmental issues affecting the well-being of kangaroos in Australia

Habitat loss and fragmentation

Over the late nineteenth century the native vegetation over large areas of southern Australia was cleared. This continued throughout the twentieth century, when northerly extensions of the land clearing became more prevalent. Over the last 30 years clearing in the southern corners of the country have been associated with 'cleaning up' small remnant areas and 'tidying up' areas for more 'efficient' farming practices. Elsewhere there has been extensive clearance of native vegetation to accommodate broadacre farming.

Land clearances and extensive tillage have resulted in the obvious loss of fertile topsoil through wind and water erosion. However, some harmful changes have been and continue to be subtle, such as the lack of regeneration of native bush species, often through overgrazing. In many areas, plant diseases such as 'dieback' (plant death due to fungi such as *Phytophthora* spp., which are distributed by human activities) have taken out susceptible plant species, leaving a highly modified plant community behind. Other factors that appear to be innocuous, such as the removal of firewood, actually remove secure refugia for animals such as bridled nailtail wallabies that shelter in hollow logs during daylight hours. Even so, the disturbance and change of available pasture species has been beneficial to a few of the larger kangaroo species.

Unfortunately, much of Australia's native bush, scrub and grasslands have been reduced to fragments that are too small to support viable populations of some vertebrate species. While the role of good-quality vegetation corridors linking habitats has been recognised as being critical to the movement of some species, all too often these corridors are inadequate and the areas set aside fail to support a viable population. Over recent decades, throughout Australia many concerned groups of people have attempted to address this by revegetating corridors to improve their effectiveness at linking isolated animal populations. In some cases this has been successful, but in others less so.

Summary

- European farming practices used for more than a century in Australia are not in harmony with the Australian environment.
- Extensive clearing of land throughout Australia has been a major factor in the decline of many kangaroo species. It has also resulted in the extinction of two species, as well as the local and continental extinction of many others.
- The fragmentation of the land with inadequate connecting corridors isolates many kangaroo colonies, resulting in inbreeding depression and restricted gene flow.

Introduced predators

Predation is an ongoing stressor to most macropods, no matter what their size. Here the smaller species and young animals, particularly those just exiting the pouch, are most at risk. Mammals predominate in the diets of dingoes, foxes, feral dogs and feral cats. Dingoes generally rely upon larger prey; foxes and dogs, on small to medium-sized prey; while cats are restricted to small prey. However, such broad generalities do not always hold. It appears that during good seasons dingoes have small pack sizes that hunt smaller prey, but during the hard times pack sizes are larger and more often larger prey is sought.

Historically, throughout most of mainland Australia equilibrium between Aboriginal people, dingoes and kangaroos was established long ago. However, the balance was disrupted by the introduction of the European red fox and domestic cat, both of which have had significant roles in the decline of many native animal species. In some instances it may be the fox that is more to blame; in other instances it is the cat. Both have caused and continue to cause havoc.

Dingoes

It is believed that dingoes evolved from the grey wolf in South-east Asia, and that their domestication commenced between 6000 and 10 000 years ago. It is probable that they were introduced accidentally into northern Australia by Asian seafarers about 3500–4000 years before the present era, and had remained isolated from other dog populations until recent times. Over the short period that it has been in Australia, the dingo has thrived and conquered most regions of this island continent. It is not present in Tasmania. With the development of the cattle and sheep grazing industries, dingoes quickly adapted to predating newborn calves, lambs and in some instances fully grown sheep. In some regions of Australia this has caused serious losses to individual graziers, and even forced some to change from raising sheep to raising cattle. An additional problem is that dingoes can freely crossbreed with domestic and feral dogs. The hybrid offspring of these matings have become a significant cause of losses to the sheep and cattle industries, and predation by wild dogs is now regarded as the most serious threat for livestock enterprises in Queensland.

In Queensland, where antilopine wallaroos occur, many pastoralists encourage their presence on their properties. They are rarely in large numbers, and it is believed that they lessen dingo predation of calves. Other than avoidance and fleeing, these macropods have ineffective anti-predator behaviours. Consequently, dingoes and wild dogs are more likely to predate a young antilopine wallaroo than attempt to predate a well-protected calf. Pastoralists in many other regions of Australia have similar views on the value of having low numbers of kangaroos to be preferred food items for the resident dingoes or wild dogs rather than their sheep or cattle. In addition, some pastoralists express the view that it is preferable to have a resident pair of dingoes or even more on their property to keep fox numbers down. Although this is unproven, it appears that dingoes are successful predators of foxes.

An average dingo weighs about 15 kg, but individuals of up to 24 kg have been measured. From a dingo's point of view, a small prey item is one weighing up to 0.5 kg, a medium size is between 0.5 and 15 kg and large is more than 15 kg. Small to medium-size prey can be caught successfully by a lone dingo whereas larger prey cannot. Across Australia, dingoes particularly seek red kangaroos, euros, swamp wallabies and agile wallabies. Juveniles and smaller adult females of the larger species are most commonly captured.

Dingoes generally work in groups to bail up and kill large kangaroos. In these pursuits they use several strategies. One is for a lead dingo to make the kangaroo change direction into the path of other pursuing dingoes. If necessary, this process is repeated until the tiring kangaroo is bailed up and then killed. A second method is to change the pursuing lead dingo regularly to maintain pressure on the kangaroo until it is exhausted and can then be bailed up. Once bailed up, other members of the pack attack the kangaroo repeatedly, generally attempting to tear the hamstring tendons of the hind legs and then rip the prey's throat out. Open terrain is conducive to successful hunting by dingo packs. Pursuits in areas of broken terrain or over moderate to dense ground cover are most likely to be unsuccessful. Smaller prey are run down and bitten along their back and neck until they collapse. Single dingoes use the same techniques on their prey but with less success.

Dingoes are primarily either predators or scavengers of carrion. The use of poison baiting to control dingo numbers could be acting as a selection agent against the carrion class.

Domestic dogs

Feral domestic dogs and household pets are significant predators of kangaroos, especially small ones. Feral dogs hunt similarly to dingoes. However, they appear to be more likely to hunt for 'pleasure'. Although household dogs kill few macropods, where human habitation with its complement of dogs is close to the preferred habitat of some of the smaller kangaroo species, ongoing harassment can seriously disrupt their colonies. An example is the harassment by dogs of the endangered Proserpine rock wallaby colonies near human habitation in the Airlie Beach region of central coastal Queensland.

European red foxes

The European red fox was first introduced into Australia in the 1850s, and had become established in the wild by the 1870s. Anecdotal and scientific evidence of fox predation of Australian native fauna

Figure 3.1: A European red fox feeding off a red kangaroo road kill. (Photo: Jiri Lochman.)

abounds. In the 1980s the pioneering field studies of Jack Kinnear showed clearly that foxes dramatically reduced black-footed rock wallaby numbers at isolated rocky sites throughout the wheatbelt of Western Australia. As a consequence, fox baiting was introduced, using baits made from dried meat injected with 1080 poison. These were usually spread over large areas using aerial distribution, and have been extremely successful in the jarrah forest of south-western Western Australia. The timing of baiting is most effective when vixens are rearing cubs and when juvenile foxes are dispersing. Unfortunately, there is evidence in the arid zone that when fox numbers are reduced, then cat numbers rise.

The continuance of this fox-baiting program over many decades has allowed black-footed rock wallaby numbers to recover and flourish, starting either from remnant populations or following reintroductions by Western Australia's Department of Environment and Conservation.

An alternative to poisoning for the control of foxes is to provide a bounty for every dead fox. Under such a system, once foxes become scarce the high costs and energy expenditure to kill the last individuals becomes more and more unlikely. Bounty systems have not been successful in Australia.

Feral domestic cats

DNA studies indicate that Australia's domestic cats were introduced with the early wave of European settlement and that they followed settlers and had established themselves in the wild across Australia by the 1890s. Because of their ability to survive and prosper in most Australian environments, they have a significant effect on any small mammal species wherever they are found. As there was no analogous hunting species in Australia before they arrived, native species had not evolved any avoidance strategies. Domestic cats are exceedingly successful killers of small macropods.

Historically the small kangaroo species such as the burrowing bettong, rufous hare wallaby and banded hare wallaby were widespread on mainland Australia. Cats and foxes are strongly implicated in the extinction of these species from the mainland. Currently, naturally occurring populations of these species are restricted to a few offshore islands of western and north-western Western Australia, where they survive because of the absence of feral predators. When these kangaroo species have been reintroduced onto the mainland, predation by feral domestic cats has been significant. Consequently predator control, especially that of cats, is critical to the success of these reintroductions. While it is distasteful and vexing to some Australians to kill feral domestic cats so that native Australia fauna may live, feral cat control is essential if our small native fauna is to survive.

A serious complicating factor in any efforts to control cat predation is that domestic cats are responsible for much adventitious predation of small species in periurban environments. Control efforts may have an impact on these individuals as well as truly feral cats, which may lead to an increase of community resistance to such programs. Lethal control programs, for cats in particular, would seem to be only possible in pastoral or remote areas.

Summary

- Cats and foxes have been significant factors in many small kangaroo species becoming vulnerable to extinction.
- Any proposed reintroductions of small kangaroo species into new or historic sites must include the management and elimination of dingoes, wild dogs, foxes and cats for these to be a success.

Weeds

There are many thousands of weed species now afflicting the Australian landscape to some degree. These can be classed as environmental, pasture and horticultural weeds, depending upon which aspect of the environment they have invaded. Environmental weeds are plants that have entered a native ecosystem and modified the indigenous plant communities and thus the ecosystem's function. These may be grasses, creepers, bushes or trees. Most weeds are introduced from beyond Australia, but some are native species that have been introduced to areas beyond their normal distribution and become feral.

Over most of Australia, and in the northern areas in particular, feral pasture plants have become weeds. In excess of 90 per cent of introduced pasture species have become feral, and many have become serious weed problems. Introduced grass species, such as buffel, mission and gamba, have become rampant and are altering the ecosystems of northern Australia significantly.

Similar examples can be given for many areas of Australia where an introduced weed now out-competes the native vegetation to the detriment of the native plant and animal communities. In a few cases, some weeds such as lantana and blackberry have provided significant daytime refugia where animals can lie up in secure sites, lessening the possibility of predation. Unfortunately such thickets do not deter cats. Some weeds do provide good food resources to native herbivores.

Summary

- Introduced and feral indigenous weeds have become extremely significant environmental problems facing Australia today.
- Weeds now occur in virtually every Australian ecosystem.
- Some weeds have significantly altered fire behaviours, resulting in a complex, diverse set of environmental outcomes.
- The impact of weeds on kangaroo species is heterogeneous, depending on the requirements of the kangaroo species involved and the nature of the weed. Some weeds provide refugia and food while others destroy important aspects of the native habitat essential for the long-term survival of some kangaroo species.

INTRODUCED HERBIVORES

There are 18 species of introduced mammalian herbivores in Australia. These are the European rabbit, brown hare, horse, donkey, pig, one-humped camel, swamp buffalo, European cattle, zebu, Bali banteng, goat, sheep, fallow deer, red deer, rusa deer, sambar deer, chital deer and hog deer. Of these the rabbit, horse, donkey, pig, camel, goat, sheep and cattle are widely distributed. They have degraded and continue to degrade Australia's plant communities. Their continuous grazing pressure across most of Australia has profoundly altered the structure of plant communities. Hence they have affected the abundance and distribution of the herbivorous marsupials.

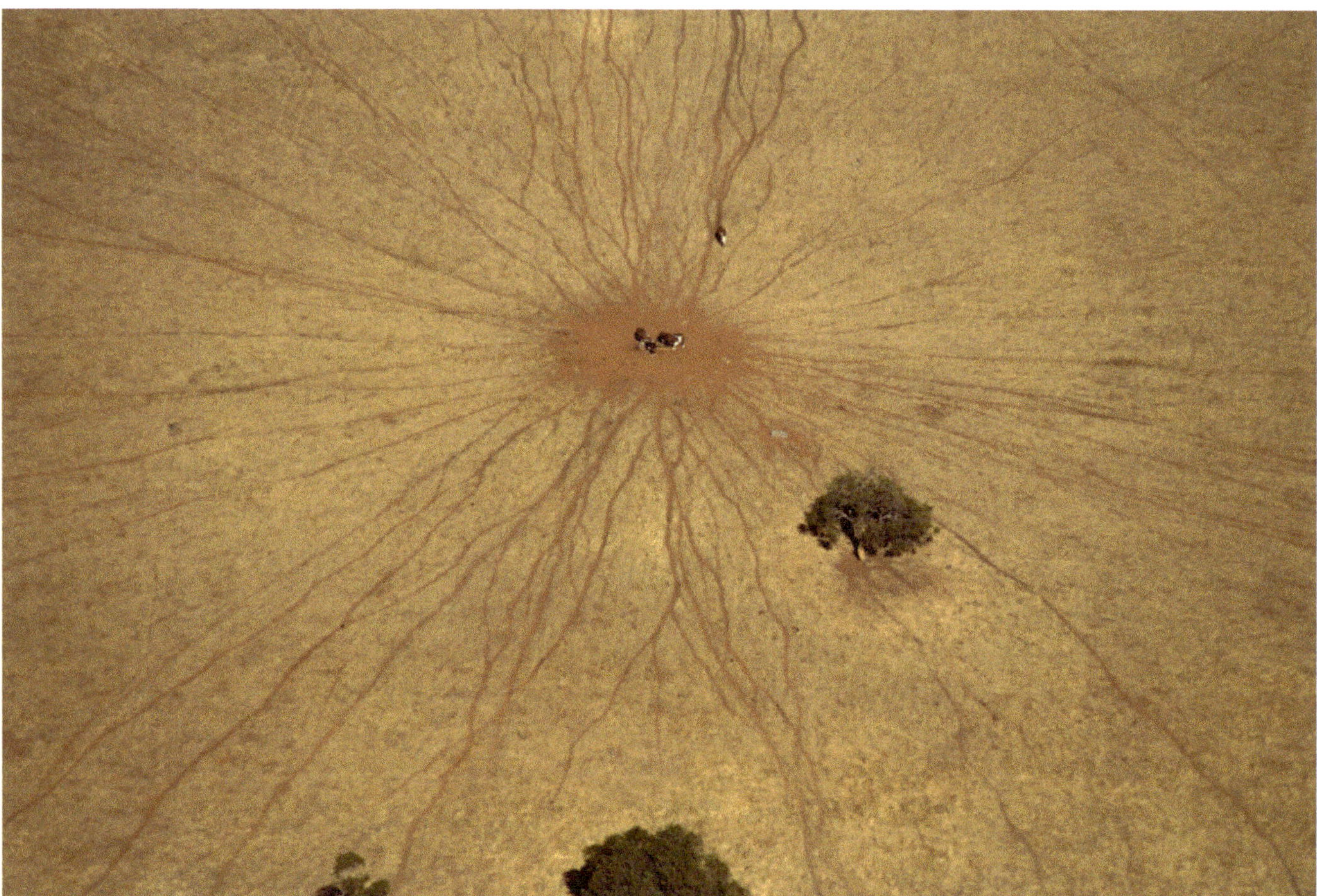

Figure 3.2: Aerial view of an area overgrazed by cattle. The cattle have eaten out all vegetation close to their principal waterer. Their radiating 'pads' lead to grazing grounds far distant to the waterer. (Photo: Jiri Lochman.)

Control of feral horses, donkeys and camels has become difficult because of public disapproval of conventional culling using firearms. Alternate control programs are becoming increasingly costly.

Summary

- In the evaluation and management of herbivores, the total grazing pressure on the environment is critical.
- Overgrazing by introduced herbivores is an ongoing significant reality over much of Australia.
- Overgrazing by kangaroos does occur in some regions, particularly adjacent to water sources. When coupled with competition for forage with feral and/or domestic herbivores, overgrazing can be extreme.
- Even though feral herbivores cause incredibly high levels of environmental damage, control of them is becoming increasingly difficult and costly.

Fire

Across large areas of pastoral Australia one can see evidence of fire. Clearly fire has, and continues to have, a major impact on the Australian environment. The biogeography of today is the result of ongoing changes over many millennia that have sculpted the landmass as well as its floral and faunal communities. The result is an aged flat landscape dominated by eucalypt and acacia woodlands as well as native grasslands. Following many years of research into the effects of bushfires in Australia, Professor David Bowman proposed that the extant floral and faunal communities are the result of three separate ages of fire. The initial age was the long period before mankind inhabited this continent. This was when lightning started major fires that burnt large areas. Over the millennia, most of the continent became colonised by fire-adapted plant communities. The second period began when the Aboriginal people came to Australia roughly 60 000 years ago. These people used fire to manage their landscape. They used

Figure 3.3: Spinifex fire being lit by an Aboriginal group in the Great Sandy Desert. (Photo: Jiri Lochman.)

frequent small burns that created a mosaic of different staged plant and animal communities. The third period commenced shortly after Europeans came to Australia. They quickly supplanted the Aboriginal people over most of the continent and discarded their fire regimens. The native plant communities altered in response to the changes in fire practices as well as the introduction of many new plant species and the introduction of domestic herbivores. Europeans saw fire in the landscape as inherently destructive and attempted to tame fire by reducing its frequency, which unfortunately meant that fuel loads rose. This, combined with climate variations such as the drier weather associated with 'El Niño', meant that when the inevitable fires did come, they were larger and often hotter than those to which the habitat had become adapted. Fire became feral. Today more than 80 per cent of Australians live in coastal cities, and to them bushfires are fascinating phenomena to be feared. Fire is dramatised and little understood.

The effects of an individual fire on the flora and fauna is influenced by many factors, such as where it is, the geology of the site, the fuel type, timing, local weather conditions, the intensity and frequency. Following fire the successional changes in the plant communities have a major effect on how the herbivores utilise the areas. Some macropods are favoured by fire. Shortly after a fire, highly mobile species such as the red kangaroo and eastern grey kangaroo move into the area to take advantage of a flush of fresh grass shoots (green pick, fresh pick). In another situation such as in the moist sclerophyllous forests, following fire there is a successional increase in truffles. Here, there is a gradual increase to a maximum of truffle fruiting over the initial 3 years post fire followed by a rapid decline, so much so that by about the sixth year post-fire senescence has set in. Hence to burn every 5 years in some situations could be advantageous to mycophagous macropods such as bettongs and potoroos.

Snapshots of the fire regimens and their effects across northern Australia

Pilbara

Historically, mulga (acacia) scrub covered much of the Pilbara. It is very fire sensitive and has a regeneration time (seed to seed) of about 20 years. Mulga on the flats has an understorey of soft annual grasses. By the dry period these have dried off and either been eaten or blown away. Effectively there is no understorey. Hence there is no continuous fuel load for any fire to run through in the depths of the mulga. Because the inner core of the Pilbara still has good stands of mulga it is unlikely that the Aborigines patch burnt there. This in turn suggests that Aborigines did not spend a lot of time there, an observation in accord with the low abundance of edible plants and animals in the mulga assemblages.

Between the 1980s and 1990s the size and frequency of bushfires throughout the Pilbara increased dramatically. By the early twenty-first century most of the Pilbara is being burnt excessively, and in places even annually. Consequently, over much of the Pilbara there has been a slow, subtle succession from mulga woodland to acacia shrubland (fire-resistant disturbance opportunists) and to spinifex grasslands. A case in point is the Karijini National Park, on the eastern side of the Pilbara, where there were vigorous stands of mulga, many recently lost to fire. Generally in this rugged landscape, fire on gentle slopes burns slowly and has little effect on the mulga. However, fire on steep slopes (1:4) burns hard and fast, resulting in the mulga being burnt off the slopes and replaced by spinifex grasslands. On the steep slopes mulga, figs and pines only remain in sites protected from fire.

Over recent times the floral changes in the Pilbara have dramatically affected the vertebrate fauna, but have had little effect on the few kangaroo species found there. In most instances the resident euros and red kangaroos move away from the path of fires and return when the post-fire 'green pick' appears. Euros and rock wallabies usually flee to high rocky ground and thus avoid the immediate impact of fire. These animals then survive on plants growing within the rocky areas, and only venture beyond the base of their rocky territories when the vegetation there has regrown.

Kimberley

From the 1990s until the present there has been a dramatic increase in the frequency and extent of the use of fire. Contrary to the suggestion that the increase in tourist numbers has caused the fires, it is local pastoralists and local Aboriginal people

Figure 3.4: Typical stand of mulga. (Photo: Marie Lochman.)

who light most of them. In many cases fires are lit indiscriminately and inappropriately.

Up until the early 1990s there was little fire used in the management of pastoral leases in the Kimberley. The recent excessive use of fire by many pastoralists to manage their properties has been an unfortunate development. All too often fires are too big, too hot, too late in the season and too frequent (annual). These fire regimens often result in an overstorey of eucalypts over annual grasslands. Perennial grasslands are decreasing dramatically in area and diversity. Obligate seeders are declining simply because the available seeds in the seed banks following hot fires are becoming fewer and fewer with each fire. In many areas spinifex is being replaced by annual grasses usually introduced pastures that have, in many areas, been classified as weeds.

A saviour for many areas of the Kimberley is its extremely rugged topography, which often breaks up large fires leaving refugia, thus creating mosaic effects. Because macropods are highly mobile, they are the least affected of the Kimberley's mammals. Where there are accumulations of large rocks on or forming outcrops there is less fuel for fire and more probability of plants surviving. Following fires, those rock wallabies that survive rely on the remaining vegetation in among their rock piles for food.

The Top End

Management of fire within Northern Territory's national parks and nature reserves is often problematic, and illustrates many of the issues that arise with the use and control of fire in the publicly controlled and owned areas. Management plans can give rise to conflicting objectives, thereby compromising the long-term success of the national parks and nature reserves. On the one hand management needs to minimise the possibility of feral fires by reducing 'fuel' levels; on the other hand there is the need to sustainably maintain biodiversity. It appears that national parks and nature reserves across the top of Australia have similar problems: too few staff and not enough money. Of necessity, the parks and reserves managers have prescribed burning regimens. Most national parks and nature reserves within the subtropics burn their perimeter at the end

Figure 3.5: A hot, vigorous fire in open woodland of the Northern Territory. These fires can reach high into the tree canopy. (Photo: Dennis Sarson.)

of the wet and then try internal mosaic burns. But they also have to cope with the inevitable feral bushfires.

A typical example is Gregory National Park, a large, mostly inaccessible park in western Northern Territory that has about 25 per cent of its area prescribed for burning each year. In some years this objective is reached, in others not. Park staff attempt to burn selectively to create patches. These account for chronology, particular sites, habitat protection and making firebreaks. In some instances they use fire as a tool to reduce the spread of some noxious plants, such as using early wet season burns to reduce areas of spear grass *Heteropogon contortus* before it seeds. However, in most years roughly another 25 per cent of the park is burnt by feral bushfires. Consequently, in most years between 20 and 70 per cent of the park is burnt. Over a decade the entire park is burnt at some time and some areas are burnt repeatedly. Many of the feral bushfires occur at a less than optimal time of the year, such as towards the end of the dry season. These fires are often excessively hot and have serious negative consequences, such as damage to the crowns of trees, and the burning of the perennial seed banks, reducing species diversity.

In other national parks and nature reserves of the Top End and over large areas of pastoral leases, introduced pasture species such as mission grass *Pennisetum polystachion* and gamba grass *Andropogon gayanus* have become weeds, and are significant contributors to the increasing prevalence of late dry-season, hot–extremely hot bushfires. Mission grass is a vigorous, heavy-seeding perennial grass that remains green late into the dry season. It provides significant fuel loads that facilitate late dry season fires. Gamba grass is a very tall (to 5 m), fast-growing exotic grass that can produce up to 10 times the biomass of native grasses, and is particularly prone to burn late in the dry season. Even though the eucalypts throughout the woodlands of the Top End are well adapted to fire, the combination of very large fuel loads (hence a longer burn in the immediate area), the greater heat produced and the effects of heat higher up the tree trunks late in the dry season kills many eucalypts. It is quite possible that frequent late dry season hot fires fuelled by exotic grasses could gradually eliminate the woodlands and forests throughout much of northern Australia. This would have a dramatic effect on the native fauna of the region.

Elsewhere the ongoing spread of a suite of other introduced pasture grasses is having a dramatic effect across much of the Australian landscape. A classic example is buffel grass *Cenchrus ciliaris*, a grass originating from southern Asia and eastern Africa that is now found across much of northern Australia. Buffel grass prefers the flatter lands, but not the cracking clays where the native Mitchell grass *Astrebla* spp. is best suited. When it is green buffel grass is highly palatable to stock, and even though it is less palatable when it dries out it is still an important dry season forage. This species recovers quickly following fire; it out-competes its native competitors and suppresses the germination of native plant species. This plant is becoming a real threat to biodiversity in the arid zones. In many areas it is considered to be a significant weed. Its impact is increasing as it gets stronger and stronger footholds. Buffel grass is slowly spreading up slopes and bringing hotter fires. These have increased the mortality of iconic trees, such as the kurrajong *Brachychiton* spp. and lancewood *Acacia shirleyi*, that rely on their seed banks for regeneration.

Savannah lands and Cape York

Suppression of fires in these areas has greatly affected many native plant communities. In some areas the terrain has become impenetrable because of post-fire regeneration of woody plants at the expense of biodiversity. Historically, in these areas regular fires burnt the tops of the woody plants so that they never grew too high or thickly. Release from this regular control by fire results in taller, bushier, woody shrubs that ultimately crowd out other plant species, particularly native grasses. Once the woody plants are above 2 m, it requires a very hot fire to top-kill them. In areas where woody plants have become unmanageable, only repeated hot fires can return them to their original normal height and density.

The antithesis of this scenario is that over large areas burning has become overly frequent, even annual, resulting in a significant degradation of the seed bank, with a resulting drop in biodiversity and hence a lessening of the quality of the habitat. Under both scenarios, antilopine wallaroo numbers and possibly northern nailtail wallaby numbers have dropped.

In this region lighting fires shortly after early storms, as long as the moisture has penetrated about 5 cm deep, is believed to be a good management tool because there is enough moisture for the native grasses to regenerate.

Arid–semi-arid zones

Cool fires in the arid zones over small areas (1–50 ha) are used successfully to patch burn spinifex grasslands. Kangaroos, because of their mobility, avoid the immediate burn and utilise nearby areas for food and shelter. Post fire there is a flush in plant regeneration resulting in abundant high-quality forage. Plant species diversity also improves. Consequently kangaroos are advantaged. In the case of extensive fires, 1000–10 000+ hectares, significant mortalities of smaller macropods occur after a fire event as a result of increased predation and loss of food resources. Although food resources recover quite rapidly, significant shelter can be removed; for example, the loss of old hummock grass clumps eliminates prime shelter sites of the spectacled hare wallaby.

In most areas, Australian pastoralists manage fire competently and cooperate willingly with staff of adjacent national parks and nature reserves. However, some pastoralists exercise poor management when it comes to the sustainable development of their properties. All too often they burn at less appropriate times of the year, causing long-term environmental damage. In many instances their decisions are driven by commercial requirements to produce a financial return to distant city owners, to satisfy corporate dictates for return on investment to shareholders, or to meet loan repayments.

Summary

- Fire is a ubiquitous presence throughout Australia.
- Since European settlement of Australia and the supplanting of small mosaic burns used by the Aboriginal people, fire has become feral.
- Agricultural practices and land ownership laws have complicated fire occurrences and behaviours throughout much of Australia.
- The introduction of 'improved' pasture grasses, domestic stock and feral herbivores have greatly modified fire behaviours throughout Australia.
- Many introduced pastures have become weeds that have significantly altered fire behaviours, resulting in a complex, diverse set of environmental outcomes ranging from the

Figure 3.6: Spinifex grassfire in the Great Sandy Desert of Western Australia. (Photo: Jiri Lochman.)

Figure 3.7: Aerial view of the patchwork appearance of small fire events in spinifex country. (Photo: Jiri Lochman.)

gradual loss of overstorey trees in some instances through to woody thickening in other cases.

- The recent fire histories of much of Australia have altered many plant communities and disadvantaged many native vertebrate species; however, few kangaroo species have been disadvantaged by fire.
- Natural climate variations, such as with El Niño and global warming, have meant harsher fires and slower than normal recovery, as well as fires affecting wider geographic areas where they would usually have had a lesser effect.
- A suite of different, area-specific management regimens is needed to protect the flora and fauna.
- Under most circumstances fires in arid and semi-arid zones result in fresh grass shoots that are sought by euros, red kangaroos and eastern grey kangaroos.
- Fire used on small areas at appropriate time intervals can be used as an important management tool. Such time intervals vary from 5 to 10 years for arid zones through to 3 to 5 years for tropical zones.
- Fire used appropriately can be advantageous to some kangaroo species.
- A few endangered kangaroo species live in small fragile habitats that are vulnerable to fire. A prime example of this is the unique vine-forest habitat of the endangered Proserpine rock wallaby, which becomes fire prone after prolonged droughts. Similarly, the restricted heathland habitat of the critically endangered Gilbert's potoroo is also threatened by fire after prolonged dry periods.

Hydrological changes

Good quality freshwater is and always has been important in determining the distribution of plant and animal communities. Changes in hydrology, such as dryland salinity, deteriorating water quality and altered stream flows, have had dramatic impacts on most of Australia's plant and animal communities.

Water sources

About 70 per cent of Australia's land mass is classified as arid–semi-arid; that is, receives 250–500 mm precipitation annually, and does not have naturally occurring permanent water but is supported by ephemeral water sources. These areas are characterised by nutritionally deficient soils (notably lacking in nitrogen and phosphorus) that support several major biomes: spinifex grasslands, Mitchell grasslands, acacia shrublands, chenopod shrublands, mallee and arid-adapted woodlands. Before European settlement these areas supported few kangaroos. The paucity of kangaroos was a result of unreliable water availability.

In the late nineteenth century and throughout the twentieth century, the introduction and ultimately widespread use of stock water troughs encouraged the geographic spread and population increases of the large grazing kangaroos (red kangaroo, eastern grey kangaroo, western grey kangaroo and common wallaroo).

Initially these troughs received water from local surface catchments and later by windmills tapping into shallow aquifers. The earliest windmills in Australia were imported from North America in the early 1800s by wealthy Queensland pastoralists. Within a few years local blacksmiths had begun to produce small windmills that were used to draw water up from hand-dug wells. For many of the early years, the water from the windmills simply filled a nearby trough, ran directly onto the ground or was directed into earthen dams. These systems were quite unreliable, because while there was wind all was fine, but once the wind ceased no water was available. It was quite a few years later that windmills were coupled with storage tanks, making the water supply much more reliable.

A second major water source, the Great Artesian Basin, formed between 230–100 mybp, lies deep under one-fifth of Australia: it occupies an estimated 1.7 m sq. km, mostly under Queensland. This holds a huge volume of water, about 64 900 m megalitres, under pressure. It has natural aquifers from which ground water flows to the surface, forming mound springs that were and still are a valuable resource supporting wildlife. The Great Artesian Basin was first tapped accidentally in 1878 when a shallow bore sunk near Bourke, New South Wales, produced flowing water. By the late 1890s drilling technology had improved, resulting in hundreds of bores being drilled throughout the arid areas of Queensland, New South Wales, South Australia and the Northern Territory across much of the Mitchell grass plains and mulga tree zones. Additional bores were drilled elsewhere in Australia in areas where other subterranean aquifers could be tapped. Thousands of kilometres of narrow, open-bore drains were used to distribute the water, particularly in western Queensland and New South Wales. These drains were convenient watering sites for domestic stock as well as for feral and native animals. The artificial water sources allowed the larger kangaroo species to move into areas where previously they had been either rare or absent. In areas where available graze was abundant the numbers of large kangaroos increased dramatically.

Figure 3.8: An old water trough in the arid zone with a merino sheep and several galahs drinking. (Photo: Jiri Lochman.)

Figure 3.9: Boiling water gushing from an artesian bore on Clifton Hills Station, Birdsville Track, South Australia. (Photo: Hans and Judy Beste.)

Figure 3.10: Untended artesian bore. The continual flow from many untended bores resulted in a lowering of the pressure within the Great Artesian Basin, resulting in the cessation of water flow from some bores. (Photo: Jiri Lochman.)

Over the second half of the twentieth century the internal pressure within the Great Artesian Basin dropped gradually, causing many bores to cease flowing.

Consequently, since the late 1990s large amounts of money have been used to cap and regulate bore flows to increase artesian pressures and improve water flows where bores are used. Controlling the bore supply has involved reducing the estimated 90 per cent wastage from open bore drains. Buried distribution pipes have now replaced many of the open drain networks. Consequently, in some areas where open bore drains previously sustained a large kangaroo population there has been a decline in kangaroo numbers, especially during and following extended periods of drought. Concurrent with this is a loss in the livelihoods of some kangaroo harvesters. This in some instances has had a negative impact on the well-being of some small isolated rural communities.

Impact of waterers in the arid–semi-arid zones

Over millennia Australia's arid–semi-arid zone biomes coevolved with marsupial herbivores. In these climatically driven areas, the number of native herbivores rose and fell as the quality and quantity of graze rose and fell. The larger kangaroo species were restricted mostly to the permanent waterholes and along the river courses. Excellent growing periods occurred spasmodically, but occasionally a run of good seasons resulted in abundant grass and standing water, allowing kangaroos numbers to rise substantially. Under these circumstances the kangaroos moved out into the surrounding areas. Once the good seasons ended, water became scarce and there was a die-off of the kangaroos. In this system the lack of surface water over much of the time meant that herbivore numbers rarely became high, and consequently the vegetation had adequate time to recover from grazing. Food availability was the principal determinant of kangaroo distribution.

Since the European colonisation of Australia the scenario in the arid–semi-arid zone has changed. Initially the changes were small and slow; however, from the mid to late twentieth century the negative effect on the environment and on many kangaroo species has been pronounced. By that time changes in land use, the effects of large numbers of sheep and cattle, as well as the effects of the spread and the increase in numbers of feral herbivores (rabbits, horses, donkeys, goats and camels) and feral carnivores (foxes and cats), were irreversible.

With favourable economic times and generous government subsidies in the 1950s, a large number of artificial water sources were brought on-line throughout the arid–semi-arid zones. Six decades later artificial sources of water are found across much of the arid–semi-arid zones. Except for the deserts (Great Sandy, Gibson, Great Victoria, Tanami and Simpson), watering points are generally less than 10 km apart. The distances vary with the pastoral practices and vegetation. Areas grazed mostly by cattle have water points about 6 km apart, but in sheep areas they are about 3 km apart. The impact of the waterers in these dry areas has been profound and complex. There is a radial symmetry in the effects of grazing around each waterer that is greatest immediately around the waterer and for up to a 1 km radius. Beyond that the impact of grazing decreases progressively.

Now at the start of the twenty-first century the effect of having much of the arid zone under rangeland grazing, coupled with the ongoing development of artificial water sources, means that most of the rangelands are under constant grazing pressure. Consequently the land does not get time to recover. Perennial native plants, notably grasses, have declined in diversity as well as in abundance, and in many areas have been replaced by unpalatable grasses and a variety of less palatable, trample-resistant shrubs. In some places, closely linked with their grazing history, there has been an increase in the abundance of hardy, perennial woody shrubs. These shrublands can become extremely dense and, on occasion, impenetrable.

In some higher rainfall areas an important but less obvious effect of the ongoing continual grazing pressure is the effect of heavy stock traffic, notably the hard-hoofed species mechanically pulverising the soil surface. This allows erosion and disrupts the nitrogen-fixation processes of algae occurring in the top 10 cm of the soil. Additionally, compaction caused by the weight of the animals, coupled with the destruction of the macropores formed by roots and burrowing invertebrates, lessens water penetration.

River systems

The majority of Australia's rivers have been damaged to such an extent that they are unrecognisable from their relatively pristine state of even 50 years ago. Farming practices, particularly land clearing, dam and weir building for irrigation and flood mitigation as well as the supply of water to huge agribusinesses producing commodities such as cotton, draining of wetlands, filling in of wetlands and the removal of ground water, have all taken their toll. Australia now has particularly unhealthy waterways. Salinity, nutrient overload, particulate matter, chemical imbalances, as well as pesticide and herbicide residues and introduced plants and animals (including fish), have made most waterways poor habitats for their native denizens. Native plants and animals are disappearing, especially freshwater fish. Many of the waterways are now clogged by exotic plant species and infested by feral fish, and their flows have often been greatly diminished by massive extraction for irrigation. This has had an impact on their safety as a source of drinking water for inland settlements. Efforts are now being made to remediate some of these problems, particularly the nitrification of waterways through excessive application of fertilisers. The inadequate and often wide fluctuations in water flows are subject to debate, with 'buy-back' schemes being favoured in some water catchments.

Although dry times are the norm for most of inland Australia, the occasional large flood can reverse many of the accumulated negative effects that occur when water is scarce and flows are limited or nonexistent. The issue is whether measures can be implemented to return rivers, as far as possible, to a state that will maintain them as corridors of life during the intervening dry periods.

In most cases the state of the rivers, swamps, marshes and lakes has little direct impact on the larger kangaroos. Unfortunately, little is known on how changes to the water availability and quality affect the smaller macropods. However, it is the overall lessening of the quality of the natural environment for which the different species have become well adapted over the millennia that is most important; and the state of Australia's rivers would appear to be largely symptomatic of that, rather than a primary cause. Many areas that were suitable for macropods even 100 years ago are now suboptimal or even unsuitable. Consequently, range distributions of many species have contracted and continue to do so. A pertinent example is the quokka, a species that prefers damper areas within the forests alongside small waterways. It has lost a considerable amount of its preferred habitat because of man-made environmental changes.

Summary

- Over the past 150 years the increased availability of fresh water through the provision of water for domestic stock has allowed significant increases in the geographic distribution of the larger species of kangaroo.
- Increased water availability has resulted in high levels of continuous grazing pressure, largely from sheep and cattle, on fragile habitats, especially in arid and semi-arid zones.
- In many areas changes in natural water flows,

decreased water quality, as well as the rise of the water table, have altered plant community diversity and productivity, to the detriment of some species of small kangaroos.

Soil salinity

Australia is an ancient continent that has had much of its large area covered by seas several times over its long history. Following each marine inundation the seas receded, leaving huge areas with large quantities of surface salts. Over subsequent millennia, rains leached much of this salt from the upper levels into the deeper parts of the soil profile. The native plants coevolved with this soil salt burden, and while salt was held deep within the soil the plants were able to flourish. The native perennial vegetation transpired large volumes of water daily and this kept ground water low. However, the massive land clearances that commenced over much of Australia in the early 1900s started an insidious march towards dryland salination over large areas of Australia. The loss of plants and the diminution of plant growth reduced the utilisation of water in the soil, so the water table rose gradually, mobilising salts stored within the soil. Consequently, salts began to accumulate closer to the soil surface, and that in turn altered the chemical profile adjacent to the plants' roots. Although Australia's native vegetation is relatively salt tolerant, excess salt is devastating. Ultimately, the composition of plant communities altered, and in many cases affected communities declined or disappeared.

The largest affected areas lie within the agricultural lands of the south-west corner of Western Australia. However, significant dryland salinity occurs in Victoria, New South Wales and South Australia. Dryland salinity is now an emerging problem in Queensland and the Northern Territory. Because the early effects of salinisation are slow and out of sight there is a long lag time, possibly up to 100 years, between the initiation of a salinity problem and it becoming obvious.

Estimates suggest that by the 2020s dryland salinity in Australia will extend over 6.5 m ha, and by 2050 between 13 and 17 m ha of agricultural land. This will affect the human population dramatically through a series of 'downstream effects', such as loss of farm viability, diminution of small town viability and damage to road and rail systems. Concurrently there

Figure 3.11: Closely associated with the removal of large tracts of forest has been the rise of the water table, bringing deep salt close to the surface. This surreal image of salination effects is a swathe of dead trees with an understorey of small salt-tolerant bushes between Narrogin and Wagin in Western Australia's wheatbelt. (Photo: Jiri Lochman.)

Figure 3.12: The effects of rapid salinity: many dead trees on the edge of Lake Taarblin between Narrogin and Wagin in Western Australia's wheatbelt. (Photo: Jiri Lochman.)

will also be significant damage to the native flora and fauna. Again, estimations of the effects on remnant native vegetation and their ecosystems suggest that by 2030 about 630 000 ha will be damaged and by 2050 up to 2 m ha will be affected. How far the salinity damage will extend into waterways, lakes and wetlands is unknown, but it is a cause for great concern. Historically, many of these areas had thriving populations of small kangaroo species, such as brush-tailed bettongs and burrowing bettongs, as well as middle-sized species, including tammar and nailtail wallabies. In most areas they have long gone, mostly because of fox predation.

Over recent decades many varied attempts have been used to alleviate increasing soil salinity. In large areas of Western Australia farmers have resorted to excavating extensive deep (2 m) drains that intersect the water table and run the saline water off their properties. This has had limited local success, but all too often simply dumps the problem onto the property of a nearby farmer or into a local creek system. In many cases the problem is aggravated by the ground water being not only saline but also acidic. In these cases the water can dissolve heavy metals and thus result in serious downstream toxic events.

Among these bleak scenarios there are success stories. Some properties have revegetated saline-affected land with saltbush-based saltland pastures. The subsequent lowering of the water table on these properties resulted in a dramatic improvement in the understorey as well as better soil stabilisation. There is increased sustainable production (wheat and sheep) coupled with greater biodiversity. Ultimately it has paid off with economic, environmental and social benefits. Such innovative risk-takers are exceptional custodians of the land and one must applaud their being in the top third of Australia's farmers.

Even with the knowledge that some enterprises have been successful, many farmers whose properties are severely affected with dryland salinity are unwilling to implement saltland pasture and / or tree-planting programs because of the cost. In addition, financial institutions themselves may be reluctant to provide funds for initiatives as they may be seen as too high a business risk. Because of the vast areas involved, current tree planting cannot keep up with the rate of land alienation due to salinisation. Overall, the problem is huge and the areas involved are vast.

The second cause of serious salinisation is irrigation. Virtually all water contains dissolved salts, such as sodium, calcium, magnesium, potassium, sulphates and chlorides, that come from the geological materials that the water has contacted. Following each irrigation event there is evaporation and transpiration of water resulting in salts accumulating in the soil. If there is adequate drainage the salts are leached away. If not, there is a gradual salt accumulation at the root level that results in poor plant growth and poor yields as well as significant alteration to the soil structure and impedance of water penetration. Irrigation can have additional side-effects, such as the accumulation of toxic elements like boron and, in some instances, severe waterlogging due to poor drainage. Like dryland salinisation, there is a suite of negative downstream effects from irrigation salinisation, in particular the increased salt loading of the water table and regional drainage systems. These in turn have negative effects on the flora and fauna. Although there has been a great deal of research undertaken into improving the water efficiency of irrigation, sustainable, practical and profitable farming techniques are still needed.

Summary

- The slow movement of large areas of agricultural land to becoming affected by high levels of salts within the upper soil profile has dramatically degraded their native plant communities.

- Remnant native vegetation adjacent to saline lands is undergoing degradation.
- Populations of native herbivores are being affected indirectly by the deterioration in the quality and quantity of available suitable forage.
- Major changes in land-use practices are essential and inevitable if the soil salinity and in some cases accompanying acidity are to be reversed and the land returned to some form of 'use'.

CLIMATE CHANGE

Between 1950 and 2005 in Australia there was an increase in the annual mean temperature of 0.9°C. Over this period eastern Australia, southern South Australia and south-west Western Australia have become significantly drier, while most of northern Australia has had a significant rise in average annual rainfall. It is suggested that by 2030 most of Australia will be between 0.4 and 2°C warmer because of anthropogenic climate change. The predicted warming will result in increased evaporation, increased summer rainfall and decreased winter–spring rainfall. In addition, climate variability will increase.

If the climate change trend of a decline in the number of individuals seen in higher altitude birds in Europe and North America is mirrored by Australian marsupials, upland species such as the northern bettong of North Queensland could be at risk. Why bird species, particularly the smaller ones at high altitude, appear to be at risk is unresolved. Factors believed to be involved are that these species have specific physiological adaptations that restrict them to those environments as well as more complex environmental issues related to the longer-term environmental changes, such as changing availability of food.

In Australia it appears that mammalian species residing and breeding in a narrow altitudinal range have little scope to move or otherwise adjust to changing temperature and moisture regimes. These small, mountainous areas are effectively sky islands. If Australia's highlands were higher, then movement up to higher altitudes could have been possible. Likewise, if they were more extensive, then corridors between similar suitable habitats and gradual movement to higher latitudes would be possible. However, this is generally not the case. One exception may be in the wet tropical highlands of northern Queensland. In this region musky rat kangaroos, endemic to the wet tropics, can be found from sea level through to the rainforests at 1000 m. This rugged region, with mountains such as Mt Bartle Frere and Mt Bellenden Ker rising to about 1600 m, provides some degree of altitudinal latitude for animals to adjust to environmental changes driven by rising temperatures. It is probable that as long as its dietary needs can be accommodated by different plant assemblages, the musky rat kangaroo has an escape route. However, in the adjacent region west of the Great Dividing Range in Far North Queensland, the northern bettong has a narrow altitudinal range of about 700–1100 m. Their normal habitat is a narrow band (less than 10 km wide) to the west of the Great Dividing Range between rainforest and dry sclerophyll forest. Specialised habitats such as this are seriously at risk to the changing rainfall levels and patterns associated with climate change.

Summary

- Climate change is altering the pattern and amount of rainfall throughout most of Australia. Initially subtle and then marked changes will follow in the distribution of plant species.
- The larger kangaroo species will be little affected.
- Small species, such as some of the bettongs and potoroos dependent on specific food resources such as 'truffles', may be seriously affected if these fungi decline in abundance or if their fruiting phenologies are altered significantly.

CONSERVATION OF AUSTRALIA'S KANGAROOS

RECENT HISTORY OF THE DECLINE OF AUSTRALIA'S KANGAROO SPECIES

Australia has an unenviable record of mammal extinctions since the First Fleet arrived in Sydney Cove in 1788. With the arrival of more and more immigrants and greatly increased human populations from the eighteenth to the twentieth centuries came significant changes to the Australian environment. Land clearing and European-style farming practices gradually spread through much of the country. Grazing pressure by domestic stock, sheep and cattle in particular, spread and increased. Concurrently there was a slow decline in native fauna and flora.

During the nineteenth century the broad-faced potoroo and the eastern hare wallaby became extinct. By the turn of the twentieth century the desert rat kangaroo, the central hare wallaby, the toolache wallaby and the crescent nailtail wallaby all became exceedingly rare, and by the 1950s and 1960s were probably extinct. Over the period from the first European colonisation until the present, many kangaroo species have had their normal distribution and population numbers significantly curtailed. This was most dramatic in the cases of the Tasmanian bettong, burrowing bettong, brush-tailed bettong, northern bettong, rufous hare wallaby, banded hare wallaby, tammar wallaby, black-striped wallaby, parma wallaby, bridled nailtail wallaby, black-footed rock wallaby, brush-tail rock wallaby, yellow-footed rock wallaby and the quokka.

Some species, such as the Tasmanian bettong, the burrowing bettong, the banded hare wallaby and recently the rufous hare wallaby, were extinguished in the wild from their mainland distributions and survive solely on a few offshore islands. Gilbert's potoroo, believed to be extinct for 128 years, was rediscovered in 1994. The rare long-footed potoroo was first discovered quite recently in the late 1960s, and the Proserpine rock wallaby was first discovered in 1976. Although the bridled nailtail wallaby was believed to be extinct, it was rediscovered in 1973. All of these small kangaroos are seriously at risk of becoming extinct in the wild over the next few decades. Fortunately state governments responsible for them have initiated ongoing recovery programs to lessen that possibility.

Determination of at-risk species

As long ago as the late nineteenth century a few concerned naturalists and wildlife experts in the northern hemisphere expressed concern over what they observed as a decline in the quality of the environment. Little action was taken, but by the mid twentieth century several national parks had been established in Europe and North America. In 1948 the Union for the Protection of Nature was formed. This small, struggling organisation, working for the betterment of wildlife, gradually grew in size to become a very large and influential body that promotes the conservation of wildlife as well as of flora and natural resources. It became the International Union for the Conservation of Nature and Natural Resources (IUCN), which is now referred to as the World Conservation Union. As a member of the World Conservation Union, Australia recognises and abides by the IUCN's system of classifying species (taxa) 'at risk' of global extinction. The IUCN has developed an objective system by which the potential risk of a species becoming extinct within a set time period can be evaluated. The various species at risk, with the basic reasons why, are listed in the IUCN's Red List of Threatened Species. Details of this can be obtained from <http://www.iucnredlist.org>. In summary, this evaluates the available data for the species (taxon) in question to determine:

a. The relative reduction in population size, over either the last 10 years or three generations, whichever is the larger
b. The geographic range of the species (occurrence and or occupancy)
c. Estimated decline in population size of mature individuals
d. Estimated population size of mature individuals; here the animal numbers are absolute, and generally many fewer than in Criteria C
e. Quantitative analysis of the probability of extinction within different timeframes.

Most of these criteria have several subcriteria, and some of these have further qualifying conditions to further categorise the species in question. Species at risk are classified as 'critically endangered', 'endangered', 'vulnerable' or 'near threatened'. From such an evaluation the species is then listed using a hierarchical alphanumeric system of criteria, subcriteria and conditions. Abbreviations used in this system are: critically endangered CR; endangered EN; vulnerable VU; near threatened NT; and least concern LC. Other criteria are: extinct EX; extinct in the wild EW; data deficient DD; and not evaluated NE.

A taxon is near threatened when it has been evaluated against the criteria but does not qualify for critically endangered, endangered or vulnerable now, but is close to qualifying for or is likely to qualify for a threatened category in the near future. In Tables 3.1 and 3.2 the features of species regarded as critically endangered, endangered or vulnerable are listed.

Table 3.1 The probability of extinction and attendant timeframe for critically endangered, endangered and vulnerable species		
Category	**Probability of extinction %**	**Time period**
Critically endangered	50	10 yrs or 3 generations
Endangered	20	20 yrs or 5 generations
Vulnerable	10	100 yrs

Table 3.2 Data for Criteria A, B, C, D and E			
Criteria	**Critically endangered**	**Endangered**	**Vulnerable**
A. Actual/projected reduction in population size	80% decline over the last 10 yrs or 3 generations	50% decline over the last 10 yrs or 3 generations	20% decline over the last 10 yrs or 3 generations
B. Extent of occurrence or area of occupancy	Occurrence <100 km^2 Occupancy <10 km^2 and any two of severe fragmentation or exists in only one location, continuing declines, extreme fluctuations	Occurrence <5000 km^2 Occupancy <500 km^2 and any two of severe fragmentation or exists at no more than 5 locations, continuing declines, extreme fluctuations	Occurrence <20000 km^2 Occupancy <2000 km^2 and any two of severe fragmentation or exists at no more than 10 locations, continuing declines, extreme fluctuations
C. Declining population numbering	<250 mature individuals and an estimated continuing decline	<2500 and an estimated continuing decline	<10 000 and an estimated continuing decline
D. Population estimated to number	<50 mature individuals	<250 mature individuals	<1000 mature individuals
E. Qualitative analysis showing the probability of extinction in the wild	At least 50% within 10 yrs or 3 generations, whichever is the longer	20% in 20 yrs or 5 generations, whichever is the longer	10% in 100 yrs

Australia's at-risk kangaroos

Within Australia there are 2 critically endangered, 5 endangered, 2 vulnerable and 11 near-threatened kangaroo species out of the 50 Australian species.

Gilbert's potoroo has the unenviable distinction of being Australia's rarest marsupial. It is critically endangered, using criterion (D) by the IUCN in 2011. This means that its total population size is estimated to be about 150 mature individuals.

The second critically endangered species is the brush-tailed bettong. This species had been listed as endangered in IUCN deliberations from 1982 until 1994. Following an extensive and successful reintroduction program they were taken off the endangered list and categorised as lower risk–conservation dependent in 1996, an extremely gratifying outcome. However, since then there has been a dramatic decline in their numbers across their range, with some local extinction and many other populations dropping to less than 10 per cent of their earlier levels. Currently it is estimated that there are only about 25 000–26 000 individuals in many different widely separated colonies. The dramatic and mysterious decline of the brush-tailed bettong's population by 90 per cent over 10 years is why they are now listed as critically endangered under criterion A4(be): the decline is a result of 'the effects of introduced taxa, hybridisation, pathogens, pollutants, competitors or parasites'.

There are five small Australian kangaroo species that were listed in 2010 by the IUCN as endangered. The species and criteria used are:

- northern bettong, EN B1ab(iii,v)+2ab(iii,v)

- long-footed potoroo, EN B1ab(v)
- banded hare wallaby, EN B1ac(iv)+2ac(iv)
- bridled nailtail wallaby, EN B1ab(iii)
- Proserpine rock wallaby, EN B1ab(iii,v).

The fact that each of these species is partially listed under criterion (EN B1) means that under criterion (B) over their geographic range, under subcriterion 1, the extent of their occurrence is estimated to be less than 5000 sq. km. Further to this, estimates indicate that at least two of the scenarios (a–c) operate:

a. Their population is severely fragmented or known to exist at no more than five locations.
b. There is a continuing decline, observed, inferred or projected, in any of the following: (i) extent of occurrence; (ii) area of occupancy; (iii) area, extent and/or quality of habitat; (iv) number of locations or subpopulations; and (v) number of mature individuals.
c. There are extreme fluctuations in any of the following: (i) extent of occurrence; (ii) area of occupancy; (iii) number of locations or subpopulations; and (iv) number of mature individuals.

In addition to sub-criterion 1, the northern bettong is also listed as being 2 ab(iii,v). Here the area of occupancy over their geographic range is less than 500 sq. km. The other criteria apply as above.

Using similar but less stringent criteria and subcriteria, the following species are listed as vulnerable to extinction:

- rufous hare wallaby
- quokka.

Likewise, the following are now listed as near threatened:

- burrowing bettong
- Tasmanian bettong
- brush-tailed rock wallaby
- black-footed rock wallaby
- monjon
- Cape York rock wallaby
- Mount Claro rock wallaby
- yellow-footed rock wallaby
- black wallaroo
- parma wallaby
- Bennett's tree kangaroo.

Genetic issues in kangaroo conservation

Even though a sound knowledge of the underlying genetics and risk analysis underpins much of the rationale for the development of recovery plans for at-risk kangaroo species, much remains to be learnt of the role of genetics in conservation biology. Many genetic issues are involved: inbreeding, loss of genetic diversity, genetic drift and accumulation of deleterious mutations. Additionally, in some cases captive breeding seems to have resulted in genetic adaptation to life in captivity that has adverse effects on reintroduction success.

In a normal large population of kangaroos there is a continuum in individual body characteristics. For example, across its geographic range the fur colour of the euro varies from sandy to rufous-red to black. The particular colour and appearance of an individual animal is its phenotype (P). The outward appearance is established by its anatomy, physiology and behaviour as determined by its genetic makeup and environmental factors. Euros also vary in many other characteristics, such as ability to tolerate high ambient temperatures; fertility rate; and resistance to disease. These continuously varying characteristics are determined by both genetic (G) and environmental (E) factors; consequently $P = G + E$. However, because of the heterogeneity of biological systems these vary somewhat, and this variance is measured as $VP = VG + VE$. Thus phenotypic variance is equal to the sum of the genetic variance plus the environmental variance.

The genetic variance is a property of the animal's inheritance. This occurs via the chromosomes through the individual genes and through their individual alleles. For each character the offspring receives one allele from each parent. However, a specific allele does not always have the same effect upon the phenotype. Its effects can be influenced by other alleles in the genotype. Within a species there may be either a large number of slightly different alleles coding for the same character or there may be few. In the former case there is strong genetic variability for that particular character; in the latter case it is poor. Usually one allele (dominant) exerts its effect over that of other alleles (recessive) coding for the same character. Many phenotypic outcomes are clouded by the interactions between the alleles in question and those from sites nearby.

Figure 3.13: A young at foot albino red-necked wallaby suckling its mother. Albino and leucistic macropods are uncommon in the wild. Inbreeding under captive situations has increased their occurrence. (Photo: Brett Dennis.)

Dominance is known to be responsible for many genetic diseases, where even a single copy of the dominant allele can cause a disease. Alternately, other diseases may be coded for on a recessive allele, and in these cases the disease only shows when the offspring receives the recessive allele from both parents. Another example of the passage of a recessive allele is that for albinism. The frequency of this recessive allele is extremely low, and it must be passed to the offspring by both parents for it to show itself. White kangaroos are rarely seen in the wild. The frequency of such events increases with inbreeding. Consequently, captive inbred colonies are more likely to have instances of albinism. Albino euros, red and grey kangaroos, red-necked wallabies and tammar wallabies are a popular focus of several tourist parks.

Some of Australia's kangaroo species are genetically depauperate because of inbreeding; that is, the mating of close relatives. This reduces their reproductive success as well as long-term survival. The degree of inbreeding is quantified by the inbreeding coefficient (F), which refers to the probability that two alleles at an organism's loci are identical by descent. When parents are totally unrelated, the offspring's inbreeding coefficient is 0. However, if there are common ancestors in an animal's lineage, then inbreeding is occurring. The more closely related the parents are, the larger the inbreeding coefficient is; for example, brother–sister, mother–son and father–daughter matings all have an inbreeding coefficient of 0.25. With half-siblings it is 0.125, and with first cousins it is 0.0625. The more distantly related the parents are, the smaller the F value. Effectively, the inbreeding coefficient is a measure of the deviation away from random mating. The inbreeding coefficient is additive through the generations. If inbreeding continues, as often occurs in captive colonies, then the F value can approach unity, resulting in the fixing of particular alleles at individual loci and increasing the genetic homozygosity of the population. Fixing is where the frequency of one allele rises higher and higher towards unity. At the same time the frequency of the other alleles at the locus decline towards zero. As inbreeding continues through the generations, the level of homozygosity increases and the level of genetic variance decreases. This has a particularly profound effect on genes or alleles carried on the X sex chromosomes that arise from the mothers. As inbreeding continues, the allele frequencies on the X-chromosome of both males and females approaches the same level. Consequently, any sex-linked characters carried by the X chromosome, including surprisingly loci for male (testicular) fitness, lose their variance.

Small population size inevitably leads to inbreeding over time. Many Australian marsupial species are inbred because for a variety of reasons their adult population size fell to a low level. Several species are now restricted to offshore islands, where in most cases populations have higher inbreeding and less genetic diversity than comparable mainland populations. An extreme example is that of the black-footed rock wallabies on Barrow Island off the coast of central Western Australia. One of Australia's premier marsupial geneticists, Dr Mark Eldridge, estimated that there have been about 1600 generations of black-footed rock wallabies over the 8000 years that Barrow Island has been isolated from the mainland. With no immigration possible, the small population has become severely inbred. The population has an inbreeding coefficient of 0.91; that is, the probability of two identical alleles at a locus in an individual is extremely high. The occurrence of no allelic diversity at a locus is 1. Their genetic diversity, as measured by microsatellite analysis, is markedly lower than the mainland Western Australian populations, such as the nearby colonies of Cape Range 200 km away, and those about 1000 km away in the southern wheatbelt.

Inbreeding leads to lowered success in reproduction and in decreased survival. These factors are referred to as inbreeding depression. In addition, inbreeding reduces the phenotypic variability available to the species, thereby reducing its range of adaptability to variation over time in features of the environment it lives in. The Barrow Island population has inbreeding depression, as indicated by a lower reproductive rate than their mainland counterparts. This is supported by the lactation frequency of adult island females being about 50 per cent compared with about 90 per cent in the mainland females. Animals from such an island source are poor candidates as founder animals for translocations because of their low genetic viability and low phenotypic adaptability.

Small population sizes are also responsible for inbreeding in the cases of Gilbert's potoroo and the Proserpine rock wallaby. In 2011, the total population size of Gilbert's potoroo is about 150 individuals. This species has probably been uncommon for a considerable period of time, and this has been exacerbated since farming in southern Western Australia cleared much of their specialised habitat. Their low genetic variance has lowered many of their fundamental functional characters, including fertility rates and juvenile survival. The great effort to halt the slide of this species towards extinction has, as of 2011, been successful and the population is currently increasing. A translocation to Bald Island has resulted in considerable breeding success.

Why the Proserpine rock wallaby lacks genetic variance is unknown. Over the past century increasing land clearance and human habitation resulted in severe fragmentation of the Proserpine rock wallaby habitat, and this has further isolated its colonies. Proserpine rock wallabies are now found in four natural populations consisting of about 25 colonies, scattered over 14 500 ha. Much of this terrain is rugged, and even 200 years ago would have limited animal dispersal because of their apparent inability to accommodate living in even slightly different rainforest habitats. These populations have severe inbreeding, which reduces their ability to evolve in the face of environmental changes. Although absolute numbers cannot be obtained because of the rugged inhospitable terrain, recent surveys suggest that the number of adult Proserpine rock wallabies is declining. Whether there is adequate genetic variance within their gene pool to cope with the predicted vegetative changes associated with climate change is yet to be seen. It is unlikely that the species will be able to adapt and survive without continuous support through human intervention.

Factors of considerable significance in the conservation of small populations of macropods include genetic diversity, genetic drift, selection, genetic bottlenecks, migration and mutations. The following summary of these issues borrows heavily from the excellent text on conservation genetics by Professor Frankham and his colleagues, Drs Ballou and Briscoe.

1. Genetic diversity is the extent of genetic variation in a population or species. This refers particularly to heterozygosity (the proportion of alleles that are heterozygous); allelic diversity (the genetic diversity within a population

measured by the average number of alleles per locus); and heritability (the ratio of the genetic variance to the total phenotypic variance).

2. Genetic drift is the change in the genetic composition of a population because of random sampling, particularly in a small population. This causes loss of genetic diversity; fluctuations in allele frequency and diversification among replicate populations as a result of accidental genetic segregation; lessened reproductive success; and lowered survival. Associated with genetic drift is the potential for loss and fixation of alleles. In a small population an allele may fall to a very low frequency and may even be lost. For example, genetic drift must be a significant factor in Western Australia, where the widely dispersed, discrete populations of the black-footed rock wallaby extend from Cape Range, to islands off Karratha, to many sites in the wheatbelt of the southern corner of Western Australia, and Esperance. Additional populations of a separate subspecies are found on islands off South Australia and in the MacDonnell Ranges of central Australia and mainland western Kimberley. Genetic drift has progressed a long way within these disparate groups, so much so that some colonies are thought to be subspecies and one or two may actually be full species.
3. Selection is the choice from a population of individuals that will survive and pass their genetic makeup on to the next generation. Although artificial selection as seen in agriculture is obvious, natural selection is much more difficult to observe. It is slower and less focused than artificial selection. Natural selection functions at all stages of the lifecycle. It manifests as the level of fertility of individual males and females, male mating success, fertilising ability of the sperm, number of offspring per female, survival of offspring to reproductive age and longevity. Natural selection also acts on other factors, such as an individual's stress resistance and endurance. The whole selection process depends on there being differences in the genetic makeup of individuals within the population. The level of selection pressure on kangaroo populations depends on the species, their life history strategies as well as on external environmental pressures. The most common selection favours the majority; that is, stabilising the phenotype and thus minimising genetic variance. A second form of selection favours a particular phenotypic extreme; that is, selection against most members of the population. This is directional selection, and is in play when selection is for characters of fitness such as arm and forearm length in male kangaroos (the longer these are, the greater the probability of an individual animal succeeding in intraspecific fights and having more successful matings). The third and least common process is disruptive selection for animals at both extremes of their phenotypic distribution. This process generally increases genetic variance. Unfortunately, natural selection among macropods has had little study.
4. Genetic bottlenecks are sudden restrictions in population size that occur either for a short time period or over a number of generations. The result of such an event is the loss of genetic diversity. The allele frequencies will differ from the original population. Recovery from a bottleneck event may be highly successful if only a few deleterious mutations remain in the population and have minimal effects in the inbreeding population. Alternatively, the deleterious alleles may become fixed, and then the population could be on the road to extinction.
5. Migration is the movement of animals away from a parent colony, and is particularly common with young males in the macropod world. Usually young females stay somewhere near their mother through most if not all of their life. Young males on the other hand are generally pressured to move away from their parents. Depending on the life history of the species this may only be hundreds of metres, or a few kilometres, or even further where they join another colony. An exception to this strategy is the black-footed rock wallaby, where single adult females with a pouch young and a blastocyst in diapause have been shown to have founded three new colonies up to 8 km from their point of origin. Such immigrants to small colonies provide essential genetic variability to a recipient colony. While it is not known absolutely how many such

immigrants are needed over each generation to remedy inbreeding depression in small colonies, it is suggested that even one individual per generation has a significant positive effect.

6. A mutation is a change in an allele or chromosome. Mutations are rare events, and very few of them are advantageous to the animal. The rate of mutation in mammals is extremely slow, on a scale of thousands to millions of years. Although mutations are the essence of evolution, in the context of a few centuries they rarely have a positive effect. However, they may have caused deleterious changes to individual alleles over the past few millions of years, and many vertebrates are believed to have deleterious alleles. These may be recessive alleles coding for specific genetic disease affecting specific physiological processes and biochemical pathways. Such disease entities become exposed when populations become small and inbreeding continues. Although such conditions are known in domestic animals such as dogs and cats, to date few such conditions have been reported in marsupials. However, as time passes and we have more exposure to captive colonies of marsupial species at risk, more diseases will be found and examined. The final factor in the extinction of some marsupial species may be newly evolving disease entities.

Animal translocation in Australia

Historically, as part of the conservation of some species at risk, small groups of animals have been captured and translocated from their normal habitat to what was believed to be a suitable habitat elsewhere. These exercises were attempts to form new self-sustaining colonies so that if the original founder populations were lost to a catastrophic event there would be a greater possibility of the species surviving. Following a number of such exercises, some of which were successful and others unsuccessful, a wealth of knowledge and experience was obtained. In 2000 this culminated in guidelines being drawn up by the Australian and New Zealand Environment and Conservation Council FaunaNet (ANZECC) on how, when and where members of at-risk species may be translocated to other appropriate habitats.

ANZECC defined translocations as the movements of living organisms from one area with free release into another. This includes restocking, introductions and reintroductions. Restocking is where a number of animals of a species are moved with the intention of either building up the number of individuals in an original habitat or improving genetic diversity. Reintroductions are where individuals of a particular species are released or established into a part of their original native range from where they had disappeared. Introductions are where individuals of a particular species are released or established outside of their historically known native range.

The Commonwealth, state and territory authorities use the following general principles for translocations:

1. Careful detailed planning of all translocations.
2. Release sites must be thoughtfully selected and well prepared.
3. Prior to any translocation, threatening processes must be identified. Countermeasures must be in place and be demonstrated to be successful.
4. Ongoing monitoring of the translocated species as well as of sympatric species including predators is essential.
5. Ongoing long-term management is critical.
6. There needs to be a long-term commitment by government and other involved participants to ensure that the necessary resources are available to maintain the special needs of the translocation site.

Additional guidelines for animal translocation in Australia can be found in Appendix A.

Island reintroductions

Most of Australia's offshore islands have been in contact with the mainland in recent geological time (14 000–7000 years ago). About 10 000 years ago, when the sea level was considerably lower than it is today, much of the continental shelf was exposed. At that time large modern islands such as Gloucester Island near Bowen in Queensland, Melville Island near Darwin in the Northern Territory and Barrow Island near Karratha in Western Australia were connected to the adjacent mainland, and many mammal species moved readily between what are now islands and the adjacent mainland. Over the period 10 000–7500 years ago, the sea level rose rapidly at about 6 mm per year, creating offshore islands and stranding members of a number of different species of mammal on these

Figure 3.14: Staff of the Department of Environment and Climate Change, New South Wales releasing a brush-tailed rock wallaby from its holding bag. This is part of their reintroduction program for this vulnerable species. (Photo: Ryan Collins, WWF.)

Figure 3.15: This freed brush-tailed rock wallaby is fitted with a location antenna so that its movements can be followed as it adapts to its new home. (Photo: Ryan Collins, WWF.)

islands. Over subsequent millennia, genetic isolation caused decreased genetic variability. Different island populations, especially those separated latitudinally by long distances, underwent genetic drift and natural selection. In a few cases this led to the formation of new subspecies.

In many cases Australia's offshore islands have the last remaining natural populations of some mammalian species. As such, these island refugia are extremely important and valuable. However, recent genetic studies of several potoroid and macropodid species on offshore islands created by recent sea-level rise have shown that some species are genetically depauperate. This has caused much debate about the long-term value of these animals for translocations.

A second group of islands, such as Tasmania and the Bass Strait islands, including Flinders and King, have not been in recent contact with the mainland, and are therefore referred to as 'insular'. The size of large islands such as Tasmania has lessened, but not necessarily prevented genetic changes among its macropods.

Within the last century both the Tasmanian bettong and the Tasmanian pademelon were found on the Australian mainland. However, now both are found only in Tasmania and on a few of its nearby islands. The Tasmanian bettong is relatively common, but is considered to be vulnerable because its prime habitat, dry grassy forests on infertile soils, is under threat from both agricultural and forestry pursuits. Similarly, the larger Tasmanian pademelon is common. It appears to have benefited from European settlement, and in some areas is superabundant and regarded as a pest species. The recent illegal introduction of the European red fox into Tasmania could be disastrous for its small potoroids and macropodids.

Fox and feral cat predation of small mainland species has caused mainland extinctions in several instances. A classic example is that by the 1940s the mainland populations of the previously widespread burrowing bettong had become extinct, and the only remaining populations were on the Western Australian islands of Bernier, Dorre, Boodie and Barrow. Likewise, the banded hare wallaby had become extinct on the mainland and was restricted to Bernier and Dorre islands. At that time, except for a few small remnant populations in the Tanami Desert, the principal viable populations of the rufous hare wallaby were on Bernier and Dorre islands. By 1992 all mainland rufous hare wallabies were extinct in the wild.

Equally, with some kangaroo species, island populations are vitally important where mainland populations are small and under threat. The quokka is a good example. Relatively large numbers of quokkas are found on Rottnest (8000–10 000) and Bald (200–600) islands off the western and southern coast of Western Australia. However, scattered throughout the south-western mainland of Western Australia, mostly in the jarrah forest and near freshwater streams, are many small colonies of quokkas, estimated to total between 3000 and 7000 individuals. Significantly, the mainland quokkas have much greater genetic diversity than their island counterparts.

Over recent times islands have been viewed favourably by government authorities as potential sites for the translocation of threatened mammal species. Appropriate islands can be used to create additional populations of a threatened species to guard against extinction in the case of a catastrophic event. Successful examples of this have been the introduction of the rufous hare wallaby to Trimouille Island, the reintroduction of the burrowing bettong (Aboriginal name, *Boodie*) to Boodie Island, and the introduction of the Proserpine rock wallaby to Hayman Island. In 2010, spectacled hare wallabies were reintroduced from Barrow Island to Hermite Island in the Montebello Island group, where the native population had been extinguished by feral cats some time after 1912.

Another potential use of islands is to introduce onto them a mix of animals from different founder sources. The subsequent population would then evolve with a unique genetic mix, and thus have a different and possibly enhanced capacity to cope with stochastic events. This is a controversial idea that to date has not been used with macropods.

Although islands have become significant natural refugia for many native mammals, the introductions of exotic mammals, including predators such as the cat, over the past 200 years onto many islands has caused the extinction or near extinction of many of their native mammal populations. Exotic herbivores (sheep, cattle, goats, rabbits and pigs) have often caused significant soil erosion and, in some instances, compaction as well as degradation of the native vegetation. The resulting poorer habitat can lower the viability of the island macropodid colonies.

However, occasionally macropodids cause serious adverse changes such as overgrazing, soil erosion and lowered plant recruitment when they have been introduced onto islands. A case in point is the longstanding accidental experiment by Governor Sir George Grey who in the late 1860s initiated the introduction of parma wallabies, tammar wallabies, brush-tail rock wallabies and swamp wallabies onto Kawau Island in New Zealand. All four species became well established and ultimately overpopulated the island, causing severe erosion and disruption to the native vegetation. By the late 1980s–1990s little of the native forest understorey remained, because macropod foraging prevented its regeneration. The leaf litter habitat had been decimated; consequently, it could no longer sustain populations of the vulnerable, native flightless birds, the Weka *Gallirallus australis greyi* and North Island Kiwi *Apteryx mantelli*. Recently a relocation and culling program was implemented to reduce the number of wallabies on the island to allow the indigenous plant and associated animal communities to return to their former pristine state. Wallaby numbers have plummeted and plant regrowth is advanced. Many bird species have had significant increases in their populations.

Case study 1: The Heirisson Prong Peninsula Reintroduction Experiment: Success, Calamity and Recovery

In 1989 community members of Useless Loop, a small mining town at Shark Bay in Western Australia, suggested that the nearby Heirisson Prong (26°4′S, 113°18′E) would be an appropriate location for the reintroduction of native fauna from nearby Dorre Island back to the mainland. At the time such interventions were in their infancy.

Heirisson Prong, a narrow, long peninsula jutting into Shark Bay, has the following positive attributes:

- It is geographically close to Dorre Island, the source of the founder animals; consequently there were unlikely to be any unforeseen climatic factors that might curtail successful reintroductions.
- Heirisson Prong has vegetation similar to that on Dorre Island.
- The geological substrate of the Prong is very similar to that of Dorre Island.
- Although the site is remote, it is served by a road that under most circumstances is accessible and is supported by a mining company.
- The narrow neck of Heirisson Prong reduced the extent of predator-proof fencing needed to isolate the introduction area. This meant a considerable cost saving.

Figure 3.16: Burrowing bettong foraging among sparce vegetation on Heirisson Prong, Shark Bay. (Photo: Jiri Lochman.)

Extensive field studies of burrowing bettongs on their island refugia (Dorre and Bernier islands) determined their basic ecological requirements and population dynamics as well as identified threatening processes. In 1992, the Western Australian Government's Department of Conservation and Land Management gave permission for the CSIRO's Division of Wildlife and Ecology to use Heirisson Prong to trial reintroductions of burrowing bettongs to mainland Australia.

Before the initial reintroductions of burrowing bettong, two criteria were set to determine the success or otherwise of the relocation program. The first was that within 5 years the population of burrowing bettongs should consist of at least 265 animals, a statistically derived number based on the reintroduced population's ability to sustainably withstand the presumed maximal level of natural predation. The second was that the population should remain at or above 265 for more than 5 years.

Process

The designated area was partitioned into a northern core area of about 12 sq. km and a southern buffer zone of about 200 sq. km.

The core area was to have all predators removed before burrowing bettongs were reintroduced, and the adjacent buffer zone was to have regular predator control so that external predator pressure on the core area was minimised. A predator-proof barrier fence was built on the southern boundary of the core area, separating it from the southern buffer zone. The initial barrier fence was 1.6 m high, and had a wire rabbit mesh to 1.2 m and a ground level horizontal mesh skirt extending outwards. An outwardly directed overhang at 45° was topped with two 7000-volt electrified wires.

In 1991 and 1992 the core area was baited 16 times using a variety of methods to present 1080 poison. This eliminated the foxes but had little effect on cat numbers. The southern buffer zone was also baited with 1080 baits and this has continued biannually ever since. The timing of baiting was found to be most effective when vixens were rearing cubs (August–September) and when juvenile foxes were dispersing (January–February). Poisoning rabbits to reduce competition for food within the core area in 1992 and 1993 was not a long-term success. However, it did have the secondary effect of reducing cat numbers.

Two predator-proof *in situ* breeding yards were built within the core area to house the initial founder burrowing bettongs. Eight female and four male burrowing bettongs were captured from Dorre Island in May 1992 and transported by helicopter to the release site. Each animal had its body characteristics measured and health indices noted before it was ear-tagged and released into small enclosures where it was held for 3 days before release into the main breeding yards (4 ha). Water was supplied, and the burrowing bettongs had access to limited amounts of supplementary food (dry dog pellets).

In September 1993 a further 18 burrowing bettongs (4 male, 14 female) were translocated from Dorre Island to Heirisson Prong. Of these, 10 were kept in the breeding yards and 8 were hard released into the core area. In October 1995 the last Dorre Island animals for this project (4 male, 8 female) were hard released. In 1998 the breeding yards were disbanded and all remaining animals hard released. From the initiation of translocations until the closure of the breeding yards 134 animals were released, and of these 80 were radio-collared. Early releases were eartagged, and those released later in the program were implanted with a passive transponder.

Monitoring of individual animals in the free-ranging Heirisson population was conducted using cage trapping along transect lines and at warrens. Following their capture the lengths of the head, tail and pes, and tail circumference and body weight of each animal were all measured. Their body condition was estimated using qualitative observations such as pelage condition, scarring and body fat level. A more quantitative estimate of body condition was calculated by dividing the cube root of body

Figure 3.17: Perentie swallowing a burrowing bettong at Barrow Island. Natural predators such as these perenties can have a major influence on the success or otherwise of translocations of small mammals. (Photo: Jiri Lochman.)

weight by pes length. This estimate was statistically comparable with similar data from animals measured on nearby Dorre Island. Reproductive status was estimated in females by examining their pouch for young and for the presence or absence of a lactating teat. In males testicular size was noted.

Regular monitoring of the population at large by radio tracking and spotlighting allowed estimates of dispersal, home ranges and basic population structure changes. Spotlighting also gave an estimate of predator and rabbit abundances. Radio tracking coupled with long periods of quiet, painstaking searching resulted in significant numbers of burrowing bettong carcasses being recovered and analysed. An analysis of these carcasses over the first 7 years confirmed that fox predation was the principal cause of death (64 per cent) and that cat predation (12 per cent) was less significant. The cause of death of 24 per cent of the retrieved carcasses could not be determined. Burrowing bettong numbers peaked in mid-year 2000 at an estimated 350 animals and then plummeted dramatically. This catastrophic decline was found to be due to a reinfestation of cats. Cyclone damage to the barrier fence in 1999 had allowed cats to reinvade the core zone, set up territories and breed. Concerted efforts using a variety of lures and trapping techniques were used, especially over summer periods when alternate prey was scarce. By 2003 the core area was ostensibly free of cats. However, only about 20 burrowing bettongs remained. On 30 June 2005 one of the major partners in this significant enterprise, CSIRO, withdrew from the project. This was because of a major shift in the philosophy and research directions by senior management of CSIRO away from individual and community ecological projects. Ongoing support from the mining company and intermittent Australian Government grants allowed the Useless Loop Community Biosphere Project Group Inc. to assume ongoing responsibility for the program. By early 2010 the bettong numbers had again risen to beyond 350 animals.

Lessons learnt

- A large amount of time and effort is needed to ensure the success of such an ambitious project. In this case professional staff of CSIRO's Division of Wildlife and Ecology, in particular Dr Jeff Short, a preeminent zoologist with extensive knowledge and experience in mammal translocations, was a major driver of the project. The many university students undertaking honours and doctoral research have been of vital importance, as well as large numbers of volunteers, including those from Earthwatch, in the success of the venture.
- The ongoing support of local people is essential. In this case, the people of the Useless Loop community, the mining company Shark Bay Salt Joint Venture (now Shark Bay Resources), and the Carrarang Pastoral Station have all made significant contributions to the success of this program.
- It was recognised that critical conservation enterprises like this one need both financial and people commitment extending over decades and possibly even longer.
- Patience: there may be a long time and many hurdles to overcome before success is achieved.
- Vigilance for changes in threat status such as predator incursions is vital.
- Effective ongoing predator control is essential.

Conclusions

The Heirisson Prong experiment has been a dramatic success. It was one of the first successful attempts in Australia to reintroduce marsupials derived from relict island populations back to the mainland. Even allowing for significant levels of ongoing predation, the number of burrowing bettongs rose by natural recruitment to a peak of about 350 animals. Although there was a subsequent plummet in burrowing bettong numbers, the lessons learnt from the earlier efforts have given all involved the confidence that animal numbers would return to previous levels and that a self-sustaining population would be established. Following the successes at Heirisson Prong, other reintroductions into peninsulas have been undertaken, such as at nearby Peron Peninsula in Western Australia (rufous hare wallabies and banded hare wallabies) and Venus Bay in South Australia (brush-tailed bettongs). However, the long-term success of the reintroductions probably depends on the level of genetic diversity of the founder populations, and unfortunately to date all island populations are inbred to some extent.

Case study 2: The Proserpine Rock Wallaby Recovery Plan: The management of a relict species with a mosaic of problems

Consideration of the Recovery Plan for the Proserpine rock wallaby is presented because this rock wallaby lives in an extremely limited area throughout which a relatively large human population is scattered. Its close proximity with a growing human population presents a complex set of conservation management issues.

The Proserpine rock wallaby was first discovered in 1976, and has since been the focus of concerted efforts to protect it and ensure its ongoing survival. Currently they are restricted geographically to central coastal Queensland near Proserpine, where four disjunct natural populations occur. There are three mainland areas where they are found (Conway, Dryander, Clarkes Range), which are characterised by large rocky outcrops and rock piles covered by semi-deciduous notophyll-microphyll vine forest. The fourth population is found on nearby Gloucester Island, where the population occupies rocky outcrops, but in this case within acacia woodland. This is a large rugged inshore continental island having many of the geological characteristics of the nearby mainland.

On the mainland, the rock wallabies utilise a variety of rocky assemblages, each being characterised by a large parent rock outcrop with adjacent rock falls creating favourable shelter as well as protection from predators. Primary locations all have large (0.6-plus m diameter) perched boulders. The rock wallabies live most of their lives within the scrub, where they forage opportunistically for leaves, fruits and flowers. In the dry season and times of drought, rock wallabies forage further from their normal home ranges out into household backyards, onto roadside verges and further into nearby grasslands. Drought is a serious problem in the area because local weather conditions dry the forest and make it highly vulnerable to fire. Fire can destroy isolated colonies, either by decimating the animals or by eliminating the unique habitat they are occupying. The very specific habitat requirements of this species mean that they have little ability to flourish in suboptimal habitats as refugia. Many of the 25 known naturally occurring colonies are in remote and inhospitable locations, and as such are relatively safe.

However, some of the larger colonies are near major towns such as Airlie Beach, busy roads and adjacent farmland. Their increasing proximity to human

Figure 3.18: Two Proserpine rock wallabies perched on rocks within their forest. The increasing urbanisation of the region surrounding their colonies is putting the existence of some colonies under threat. (Photo: Hans and Judy Beste.)

habitation and activities has introduced additional risk factors. Predation and colony disturbance by feral and domestic dogs and cats has increased. Cats have transmitted the highly contagious protozoal disease toxoplasmosis to some colonies, causing many rock wallaby deaths. Road mortalities have increased with the increase in urbanisation. The increasing human population in the area has led to the gradual loss of rock wallaby habitat.

An additional problem for the Proserpine rock wallabies is the presence of a relatively large number of unadorned rock wallabies in the area. These appear to compete for habitat with the Proserpine rock wallabies. The extent of this is unknown.

As a nationally listed threatened species, the Proserpine rock wallaby is the subject of a statutory Recovery Plan that must be implemented by the Queensland Parks and Wildlife Service (QPWS), under the Commonwealth *Environment Protection and Biodiversity Conservation Act* 1999. Details of how the plan is drawn up and implemented are set out in Appendix B. Basically a Recovery Plan identifies the threatened species, its national distribution and actual and potentially suitable habitats. The plan requires a study of the species' biology and all threats: predators, disease and habitat risks. Finally, it recommends remedial actions, defining the area for treatment. The initial Recovery Plan for the Proserpine rock wallaby was submitted to the Queensland Department of Environment, Brisbane in 1991. Since then there have been regular progress reports to and updates of the Recovery Plan approved by the Commonwealth and Queensland Governments. The overall objective was to increase the population size and area of distribution of the Proserpine rock wallaby within 5 years.

Under this there were five specific objectives:

- to maintain and protect known habitat and ensure that the Proserpine rock wallaby continues to exist in the wild
- to increase community awareness, in particular among those freehold landholders owning areas of habitat critical to Proserpine rock wallaby survival
- to release captive-bred Proserpine rock wallabies to the wild within suitable and appropriate habitat and ensure the population is self-sustaining in the wild
- to minimise disease, incidental kills and other destructive impacts on Proserpine rock wallaby populations
- to coordinate the implementation of the Proserpine rock wallaby Recovery Plan to enable the other objectives to be met.

To assess the success or otherwise of the specific objectives, five complementary, measurable performance criteria were identified. The performance criteria were:

- Areas of habitat are identified, mapped and protected. All known colony sites are effectively protected, and their status is well known.
- Private landholders actively support the program and the local community is aware of and participates in the recovery plan.
- The captive breeding program continues, and selected Proserpine rock wallabies are bred and released to establish successfully at least one additional self-sustaining population.
- Human impact on Proserpine rock wallaby populations is reduced to an ecologically sustainable level.
- The recovery effort is effective and efficient.

To achieve the five specific objectives, five specific actions were implemented. These were continuations of activities undertaken in earlier recovery plans. A subset of individual actions underpins each generic specific action. The main actions were:

- to continue systematic habitat surveys and monitoring, protection of habitat and monitoring of populations
- to secure landholder support, involvement and commitment to the maintenance of habitat and increase public awareness of and participation in the recovery program through the actions and directions of the recovery team
- to assess new sites and undertake release of captive-bred Proserpine rock wallabies into the wild, monitor the newly released population and maintain the captive colony
- to install effective reflectors and road signs, and promote the control and containment of domestic animals and introduced toxic plants
- to coordinate the Proserpine rock wallaby recovery team, and administer, resource and support the implementation of the Recovery Plan.

Results so far

- A recovery team has worked successfully to protect the ongoing well-being of the Proserpine rock wallaby for more than 20 years.
- Most Proserpine rock wallaby colonies appear to be safe.
- Ongoing monitoring of Proserpine rock wallaby colonies is in place.
- Acquisition of private land having Proserpine rock wallaby colonies has proceeded.
- Voluntary conservation agreements with private landholders who have Proserpine rock wallaby colonies on their properties or nearby have increased.
- Discussions between the recovery team, the Whitsunday Shire Council and land developers have developed more stringent requirements for Proserpine rock wallaby habitat protection.
- Mapping of suitable habitat is approaching completion.
- Research programs have been progressing well.
- The use of Swareflex reflectors along roads having a history of Proserpine rock wallaby mortality appears to have decreased the number of road kills.
- Microsatellite DNA analysis of scats to determine individuals and species is proceeding slowly.
- Public awareness of the Proserpine rock wallaby is rising.
- The translocation of Proserpine rock wallabies to Hayman Island has been successful. This has added a further population lessening the possible effects of mainland disasters such as fire.

Problems

- Some suitable habitat in the Proserpine–Shute Harbour area has been lost to development.
- Forty per cent of the suitable habitat available to the Proserpine rock wallaby is freehold land, and as such is not under the recovery team's control.
- The retirement and transfer of people, particularly members of the recovery team with in-depth knowledge of the Proserpine rock wallaby, are creating problems.
- Too few people in the local community know and care about the long-term fate of the Proserpine rock wallaby.
- The presence of healthy numbers of unadorned rock wallabies in the area has created a few problems. The local human population is generally unaware of their presence. Consequently they consider the two species as one, and so do not realise that the Proserpine rock wallaby is really endangered. A second problem is that it is not known whether the unadorned rock wallabies compete for habitat with the Proserpine rock wallaby.

Lessons learnt

- The adoption of the Proserpine rock wallaby as an 'icon species' by the local and tourist populations has been essential to the ongoing success of the Recovery Plan.
- The ongoing enthusiastic activity by concerned members of the community has been critical to the success of the Recovery Plan.
- Even though a great deal of effort and hard work has been undertaken in studies of the Proserpine rock wallaby, there remain huge gaps in the knowledge of their basic biology.
- The transfer of data and knowledge within a recovery team as its composition changes over time is crucial.
- Knowledge of the role of exotic disease such as toxoplasmosis in rock wallaby ecology is poor.
- The need to relocate the Proserpine rock wallaby beyond its currently known boundaries is essential.

Conclusions

- The poor genetic diversity of the Proserpine rock wallaby is a threat to the long-term conservation of this species.
- The long-term successful conservation of the Proserpine rock wallaby will be determined by the successful involvement of the local community. Public education and publicity programs are essential in the exercise.
- The introduction of Proserpine rock wallabies into novel habitats could be a key to their successful conservation.

4. Kangaroos and humans today

National attitudes to kangaroos

Australians consider kangaroos as iconic and part of their heritage. As a nation we have acknowledged their special place in our hearts by placing the red kangaroo supporting the left side of the Federation Shield on the Coat of Arms of Australia. The emu, another icon, supports the right side of the shield. Kangaroos also feature on the Coat of Arms of Western Australia, New South Wales and the Northern Territory, and a demi-kangaroo is located at the top of the Victorian Coat of Arms. Australia does not have a specific faunal emblem. However, by popular acceptance the kangaroo fulfils this role. Australians enjoy seeing kangaroos roaming free in the countryside and in many instances we have them near our homes. This is remarkable when one considers that the majority of people living in Australia live in cities and large towns. There is an intrinsic value in our being able to readily observe free-ranging kangaroos. Somehow we need to acknowledge and treasure this if we wish to have this privilege remain 'in perpetuity'. As Australia's human population continues to rise there are a number of pertinent questions that need to be considered by the Australian public:

- Do we want kangaroos roaming freely through the Australian landscape?
- How do we value kangaroos in our modern landscape?
- If we want them in our landscape, how much are we willing to pay for this?

Currently there are tens of millions of large grazing kangaroos across the Australian continent as well as many millions of the smaller species.

Tourism

Tourism is a major industry in Australia, directly providing 500 000 jobs and indirectly another 500 000 jobs across the continent. Official estimates in 2010 of the income generated by tourism range from 86 to 92 billion Australian dollars, domestic tourism accounting for approximately 73 per cent of the total tourist expenditure. The domestic revenue was generated by roughly 67 million overnight trips with expenditures totalling about $43 billion, and about 152 million day trips with expenditures totalling in excess of $15 billion. The great variety of scenic and wildlife tourist destinations encourages local residents to travel within Australia to sample the wealth of experiences available to all. Fortunately a large number of important iconic landscapes and habitats are on public lands of one form or another, and as such are protected to some extent.

To international visitors the kangaroo is the foremost image of Australia, followed by the koala. Australia is perceived internationally as being a clean, green and safe destination with a stable government and a well-developed infrastructure for visitors. Its green credentials are responsible for a significant number of the overseas visitors coming to Australia. More than 5.4 million international visitors entered Australia in 2010 and spent in excess of $23 billion in goods and services. This expenditure puts Australia at number 8 in the top 10 of international tourism receipts in the world.

Surveys show that 68 per cent of these international visitors wish to see and if possible interact with our animals. About 18 per cent of all international tourists attracted to Australia are interested primarily in our wildlife. This equates to nearly $5 billion from ecotourism with wildlife, especially kangaroos, as the most popular fauna. These economic findings support the sustainable value of wildlife viewing.

Wildlife tourists use all modes of transport – air, rail, water, road and foot – to have pleasant meaningful experiences with wildlife. Consequently they inject large amounts of money into these and allied enterprises. Tourists also consume goods and services, of which food and accommodation are a large and significant component. Cafes, restaurants, grocery shops, small bed and breakfasts, caravan parks, motels, hotels and specialised wilderness resorts supply these. In addition, there is the employment of many people in a wide range of specialised activities, including wildlife managers, guides, tourism operators and transport operators. A lucrative niche market in items such as camping equipment, guidebooks, binoculars, telescopes, video and digital cameras and appropriate clothing has grown to service the wildlife–ecotourism aspect of the industry. As part of the tourist industry, souvenirs featuring kangaroos are essential and sought after. These range from processed hides to cuddly toy kangaroos, postcards, posters, photographs and children's book on kangaroos. It is interesting to note that the children's literature has an enormous number of titles featuring kangaroos, but within the realm of adult reading there are few titles indeed.

Figure 4.1: The pull of kangaroos on the general public. Here a large number of people are watching a group of red kangaroos being fed 'ad lib' grain simply poured onto the ground. (Photo: Hans and Judy Beste.)

Figure 4.2: A tourist in Stirling Range National Park in southern Western Australia feeding wild western grey kangaroos. Bluff Knoll is prominent in the background. (Photo: Jiri Lochman.)

Overseas tourists are fascinated by the image of a kangaroo. They have seen and read the promotional literature of Australia that strongly promotes 'The Kangaroo'. When they visit Australia the most popular animals that they wish to see are kangaroos, preferably in the wild. Second, they want to interact with the kangaroos, and in particular to be photographed in close proximity to an individual or a group of kangaroos. Next they wish to feed a kangaroo. Consequently there is a demand by the tourist industry for access to tractable kangaroos. Generally this is readily accommodated. Tourist destinations with tractable animals in suitable enclosures are common throughout Australia, and especially so near all major cities.

In facilities where tourists are able to mingle with kangaroos, the kangaroos maintain their individual

Figure 4.3: An apparently totally relaxed eastern grey kangaroo with a young admirer. Most kangaroos will not tolerate such a close encounter. (Photo: Hans and Judy Beste.)

'flight distance'. If a tourist approaches the animal and moves within that distance the kangaroo generally moves away. If the person again moves towards the animal it is tolerated until again the person enters the threshold of the kangaroo's flight distance, and again the animal will move off. Different kangaroo species have different flight distances, and within a species different individuals have different flight distances. As long as humans move slowly and quietly among kangaroos there is little likelihood of problems arising. When humans make sudden rapid movement towards an individual or a group of kangaroos they will take flight, a natural response to a perceived threat. On rare occasions an individual kangaroo may exhibit aggressive behaviour to humans.

Possibly the most specific promotion of 'The Kangaroo' in Australia has been to golfers. Much of the promotional material used to attract overseas golfers to purchase lucrative package deals to Australia (flights, accommodation and access to exclusive golf courses) has photographs of a golfer with a kangaroo in the picture. Customers who take up such opportunities identify their being 'touched by nature' as they play their games of golf among the kangaroos that live on the golf courses, a situation perceived as a marvel by Asian tourists.

Australia's national airline, Qantas, has a symbolic red kangaroo on the tail of all of its aircraft. This logo is a highly recognisable identifier of Qantas worldwide.

Even though tourists want and in many instances demand to see kangaroos, most know little about them. There has been little to no attempt to educate Australians or overseas tourists to the fact that

Figure 4.4: Western grey kangaroos mingle with golfers on many Australian golf courses. This is a major drawcard for international golfing enthusiasts. (Photo: Len Stewart.)

there are a large number of kangaroo species. Most people are aware that there are one or two maybe three or four species. Australians are aware of the large grazing species, especially red kangaroos and grey kangaroos. They are also aware that there are smaller kangaroos such as wallabies, but don't know which one is which. Generally they are aware of the particular species in the geographic region with which they are most familiar, such as the quokka of Rottnest Island off Perth, Western Australia, the tammar wallabies of Kangaroo Island off Adelaide, South Australia, and the rock wallabies of various rock piles or cliffs of many of Australia's national parks and nature reserves.

Most Australians are unaware that there are nine species seriously at risk of extinction and a further 12 species near threatened. On the positive side, the local populace in specific geographic areas may be aware that they have a rare or vulnerable kangaroo species. Excellent examples are the Proserpine rock wallaby of Airlie Beach in central Queensland, and Gilbert's potoroo of the Albany area in Western Australia. In each case local publicity has overridden the blasé attitudes that most Australians have towards their fauna and flora.

Figure 4.5: A tourist group getting up close with a Quokka on Rottnest Island, an A Class Reserve, just 20 km from Fremantle harbour. (Photo: Bill Belson.)

In recent times Rootourism, a wildlife tourism information provider, has posted an informative website <www.rootourism.com> that shows tourists where they can best observe most of Australia's 50 species of kangaroo. On the website called 'The Kangaroo Trail' they provide an interactive map showing where particular species can be located, as well as a fact sheet on each species covering the identification, habitat, foraging behaviour, reproductive behaviour, social organisation and further readings about each one. This not-for-profit enterprise has and will continue to assist both casual and serious kangaroo observers in their pursuit of these animals.

Australia has the worst record of mammalian extinctions over the recent 200 or more years of any country on earth, and for several decades now children have been informed of this and of Australia's environmental issues throughout their primary and secondary school education. Many children are well informed and quite concerned about what is happening now, and what the future holds on such issues. However, these serious messages appear to be lost on the wider community. Generally, individual adults are to some extent aware of the serious demise of Australia's environment and of the possible loss of many of our mammals in the near future. However, when action or a decision is required to alleviate an environmental problem, the reaction seems to be one of apathy. As a nation we are aware that Australia is the driest continent on earth, that there is a serious and growing problem of salinisation of extensive farming and runoff areas, as well as overclearing of vegetation. Yet we allow the few who are responsible for these activities to continue unchallenged. Business interests in Australia as well as overseas continue to determine the chain of decisions and events that end in the above disasters. We Australians are renowned for our patriotism, yet we don't challenge the many and varied approaches that exploit our land

Figure 4.6: A chance encounter between mutually interested parties at Coral Bay on the mid-west coast of Western Australia. (Photo: Bill Belson.)

detrimentally. It is apparent that short-term vested interests are the current winners in most situations where there is an interplay of economic, political, and environmental and ecological issues. We have more than enough knowledge to develop a truly meaningful relationship with our land, yet most of us have not done this.

Even so, Australians take 150 million day trips annually to visit scenic sites where they often can enjoy casual encounters with kangaroos. Indeed, in many instances Australians seek out such encounters. About 50 per cent of adult Australians visit a World Heritage Area or State or National Park each year. Areas such as the Grampian Mountains of Victoria and the Blue Mountains of New South Wales are noted for their abundant, friendly and approachable kangaroos. While most people are not so focused on travelling to a site solely for the kangaroos, the presence of kangaroos moving naturally through the picnic grounds, barbecue areas and camping grounds is a delight to those who experience such situations. In tourist areas, local groups of kangaroos become acclimated to the presence of people and their paraphernalia. Consequently, in many camping grounds and city parks kangaroos move readily among humans without any display of aggression. However, they are very aware of where people are and to some extent what they are doing. In some camping grounds during the school holiday period, when the area is full of noise and bustling with active children, the resident kangaroos temporarily avoid the area only to return once the area has quietened down again.

Sometimes familiarity does breed contempt. Occasionally a large kangaroo, usually a male, may make threatening overtures to and even attack a human. Usually this consists of the animal leaning backwards, taking its weight on its tail and then stretching out its forearms attempting to grasp the person around the neck and pull him or her in towards its body. Generally this is as far as the attempt goes. Usually the person breaks off the engagement that may have resulted in severe scratches by the animal's claws. However, sometimes the attack progresses, with the kangaroo holding the person tightly and then, while balancing on its tail, it lifts up its hind legs and lashes out with its feet at the person's torso. This can result in severe lacerations and bruising to the person because of the animal's great strength and the toughness of the nails on its feet.

Living with kangaroos

Kangaroos occur in most regions of Australia, and are welcome and valued residents. Most households enjoy interacting with kangaroos. However, in some cases kangaroo foraging can result in conflicts of interest.

On a minor scale, kangaroos feeding in a person's home garden is common in many semi-rural and agricultural areas. It can be annoying, but has several readily available solutions. Exclusion fencing is generally the best one. Alternatively, noxious sprays such as chilli mixtures, ammonia and lapsong tea

Figure 4.7: Western grey kangaroos relaxing during the day in a camping ground at Cape Le Grand National Park in southern Western Australia. (Photo: Jiri Lochman.)

have all been used with varying degrees of success. A novel approach has been to blend a number of very rotten eggs in water and to spray this onto and around affected areas to repel the kangaroos. This can be highly effective.

On a larger scale, wallabies and pademelons can become significant pests of commercial crops such as cut flower and tree farms. When kangaroos are in large numbers, they can cause severe economic damage, and as such have been the focus of endeavours to mitigate their effects. In many cases, non-lethal methods, such as fencing, tree guards, egg-based repellents as well as trapping and relocation of individuals and small groups, have been effective in managing damage to commercial enterprises.

The most controversial instances of close proximity of humans and kangaroos occur in periurban regions. Here an increasing number of negative interactions between humans and kangaroos has resulted in serious debate over how to mitigate these. A current example is the gradual increase in eastern grey kangaroo numbers in and around the city of Canberra. Because of public protests over many years little had been done to limit the increase in the kangaroo population. However, the habitat around Canberra is ideal for eastern grey kangaroos, and their density has risen to an extremely high level. At very high densities their grazing pressure damages the environment severely. Additionally, there has been an increase in the number of collisions between kangaroos and motor vehicles in and around Canberra to an estimated number in excess of 5000 collisions annually. Other than the death and injury to kangaroos, these have resulted in injury to humans as well as in the considerable costs incurred in repairing vehicle damage. The Australian Capital Territory Government authority, Territory and Municipal Services, has the responsibility of resolving the problem of too many kangaroos in the area under their jurisdiction. Fertility control, management translocations and culling have all been considered in depth. However, there are significant limitations in all approaches.

Currently the methodologies and effectiveness of macropodid fertility control are still undergoing research and evaluation. Some certainly work well on individuals and for small groups in the few species studied to date. However, the human intervention necessary to administer the drugs to individual kangaroos has several potential negative welfare implications. Additionally, the need to repeat the procedure as well as cost implications detract from this being viable at present for large numbers of animals.

The principles underpinning management translocations and conservation translocations are basically the same. However, before any management translocation of animals in conflict with humans is undertaken, one has to consider whether this merely relocates the problem elsewhere. There are frequent proposals to translocate eastern grey kangaroos and western grey kangaroos because both species have significant populations living close to growing human populations. Land developers often have a need to address the issue of what will happen to the kangaroos on their land that is scheduled for development. In some cases animals can be translocated effectively, but all too often this is not possible, either because there is no suitable area available for translocation or the entire procedure is too expensive. The alternatives to translocation in these instances are limited.

Similarly with kangaroos isolated within totally enclosed areas, unless there are measures in place to limit population growth, an increase in kangaroo numbers can result in their starvation as well as in severe environmental degradation. Such situations have caused serious animal welfare problems in several fully fenced-off nature reserves and Defence lands across Australia. In most situations the destruction of the kangaroos is the most likely outcome. However, the choice of how to do this can be problematic. Chemical tranquillisers administered via dart gun followed by lethal injection have been used on several occasions. Although this may be suitable for individual animals or small groups, it is not for larger numbers. In these cases animal welfare for the kangaroos can be particularly bad. The principal alternative is to use professional shooters. This is cost effective and has few animal welfare drawbacks. However, in some cases, especially in or adjacent to built-up areas, the use of firearms is not a politically acceptable option.

In some circumstances, such as during drought and when kangaroos and domestic stock numbers exceed the total carrying capacity of an area, they damage their own habitat. A case in point is the 48 000 ha Hattah-Kulkyne National Park in north-western Victoria, where from the 1950s through to the early 1980s rabbits, sheep, goats, western grey kangaroos and to a lesser extent red kangaroos were in large numbers and denuded large areas of the park. There was extensive damage to the landscape, including scalded areas of floodplains and eroded dunes. The normal plant biodiversity had declined dramatically, particularly the woody perennials. Recruitment of slender cypress pine *Callitris gracilis* and buloke *Allocasuarina luehmannii* was severely compromised by rabbit and kangaroo overgrazing. The remaining cypress pine-buloke woodlands were mostly senescent. Damage to the soil profile was severe.

Observations and population modelling carried out by the University of Melbourne determined that the only viable resolution to this problem over this large area of western Victoria was to destock the domestic herbivores, control rabbits and implement managed culls of the kangaroos. These culls were undertaken by professional shooters under the direct supervision of senior park staff. When culling, they endeavoured to remove family groups rather than random individuals. The removal of the severe grazing pressure has resulted in a great improvement in plant biodiversity and plant densities. Cypress pine and buloke woodlands have reestablished to some extent, but will need several more decades for the plant communities to return to their former glory. This slow recovery is typical of plant communities in arid zones, as all too often it is episodic and is particularly reliant on favourable rainfall. Regular culling of kangaroos, guided by appropriate data and scientific advice, is an essential management tool for maintaining the health of parks such as Hattah-Kulkyne.

Culling

Culling is the selective or opportunistic removal of individuals or groups of animals. Historically, kangaroo culling has been allowed under permit by either shooting or poisoning. Over recent times poisoning has been abandoned throughout Australia on animal welfare grounds. All Australian states and mainland territories have destruction permits allowing the shooting of individual or small groups of kangaroos (culls) under specific circumstances. These damage mitigation licences or destruction permits are approved on a case-by-case basis to allow the shooting or in some instances alternate methods of destruction of 'problem' kangaroos.

Generally, destruction permits are issued for a defined period and for a specific number of animals. In all cases, the destruction of an animal or animals must comply with the strict conditions set in state and commonwealth legislation for the humane destruction of kangaroos. However, adherence to these conditions relies on the integrity of the permit holder.

Destruction permits allow landholders to alleviate defined unwanted impacts caused by kangaroos on their property. Across Australia these are issued where competition with livestock is deemed detrimental; where serious crop damage is occurring; or where serious environmental damage, especially to biodiversity, is happening. In these instances the kangaroos in question are considered to be pest animals. However, for any control of pest animals a case must be made of the rationale for reducing the impact of the population; for example, increasing plant biomass, lowering crop damage to an acceptable level or reversing the detrimental environmental damage.

In all states and territories the appropriate authorities manage kangaroos on gazetted parks and reserves proclaimed or set aside for conservation as well as on gazetted tenured forests. Occasionally the landowner requiring a destruction permit is one of these government departments wanting to reduce kangaroo numbers in a designated wildlife reserve, park or forest. In these public lands it is common for animal numbers to rise gradually to a stage where they can no longer obtain enough food and cause severe damage to their own environment.

Another major circumstance where destruction permits are issued is when there is a risk to human safety. This is usually for the dispatch of an individual large male kangaroo. Large male kangaroos, including castrate males, can become extremely aggressive towards humans, especially when humans have raised them and they lose their fear of humans.

Process

Culling is most often undertaken using firearms. The highly proficient marksmen who shoot the kangaroos must follow all the rules and regulations put in place by the various governmental authorities to dispatch each animal, and in particular adhere to the strict guidelines laid out in the Commonwealth Government document *Code of Practice for the Humane Shooting of Kangaroos and Wallabies for Commercial*

Purposes. However, in some cases there has been public opposition to the use of firearms to cull kangaroos, so alternatives have been used. The most common one has been to dart each individual with a tranquilliser and wait for it to quieten. The animal is then captured and given a lethal injection.

Responses to comments for and against the culling of kangaroos

Culling is controversial to many people. Below are some of the arguments for and against culling.

1. *Culling, if it has to be done, should be carried out by capture and then lethal injection.*
   Chemical tranquillisers administered via dart gun followed by lethal injection have been used on many occasions. Although this regimen may be suitable for individual animals or small groups, it is not so for larger populations. This can have serious risks to the well-being of the individual animal, ranging from the stress due to human proximity; stress due to the procedures used to herd and hold the animals within a small area; capture myopathy (critical increase in body temperature or severe lactic acidosis); inadequate chemical tranquillisation from darting; through to broken limbs.
2. *Contraception is a viable alternative to culling.*
   Immunocontraception and contraceptive implants may become useful management tools for individual and small numbers of sedentary kangaroos on urban fringes and to some extent animals within small confined areas. This is an evolving field of research where progress is being made. However, cost implications for their use and the need for regular and even annual repeat procedures detract from their widespread adoption. Wherever physical or chemical (darting) capture or restraint is used, there is always a serious risk to an animal's welfare such as capture myopathy and broken bones. Chemofertility agents do not reduce the need for culling in situations where food resources are already depleted and kangaroos are starving.
3. *Kangaroos should be relocated to an alternative place having a suitable habitat.*
   This comment is frequently used in southern Australia, mostly referring to both species of grey kangaroo. Property developers, in particular, have to address the issue of what will happen to the kangaroos on their land scheduled for development. However, in most instances there is no suitable 'vacant' habitat. Most areas already have their own population of kangaroos, which are to a greater or lesser extent in harmony with their local environment. To place other individuals into these areas can seriously disrupt the group dynamics, resulting in fighting and harm to individual animals, usually the recent introductions. In some cases the total carrying capacity of the land is already at its limit, and additional animals can tip the balance, compromising the well-being of all animals in the area. The process of relocating kangaroos, especially the large grazing species, can be extremely stressful. The close proximity of humans, and the handling, transportation and ultimate release are all contributors to stress, and in their worst case can have serious detrimental effects on animals, such as capture myopathy and broken limbs.
4. *Near human habitation kangaroos frequently cause severe damage to the environment.*
   In some instances kangaroo numbers in particular areas exceed the grazing capacity of the land, causing serious damage to the environment. This is particularly true where animals have been fenced off within and cannot move beyond an area. Under these circumstances animal numbers can rise quickly, and a great deal of damage can be done to the enclosed environment.
5. *Culling with firearms near human habitation poses a threat to public safety.*
   Although culling of kangaroos using firearms near human habitation may pose a risk to nearby humans, this is minimised by employing highly qualified professional marksmen who use appropriately modified ballistics to lessen the risk to public safety while still ensuring ethical killing.

The commercial kangaroo harvest

Ever since domestic herbivores were first introduced into Australia, kangaroos have been perceived as competitors for available pasture. Little consideration has been afforded them with regard to their intrinsic value or their value as a resource. Throughout the 1800s and early 1900s, landowners and property managers solved the 'problem' of 'too many' kangaroos by shooting and poisoning millions of both small and large species. Over this period extensive land clearance and modified ecosystems lessened the quality of the earlier landscape for the native fauna and flora. Intertwined with the dynamic interactions between kangaroos and domestic stock, the introduction of the European rabbit in 1859 added a further competitor for available pasture. Within five to six decades rabbits had spread successfully across most of the southern half of Australia and devastated most of the habitats. Rabbit numbers rose so high that commercial shooting of rabbits for their skins developed rapidly. This industry became a significant part of the Australian landscape until the myxomatosis virus was introduced into the rabbit population in 1950 to reduce their densities. Myxomatosis decimated the Australian rabbit population, resulting in the collapse of the wild rabbit harvest industry. As a consequence, many rabbiters sought alternative income streams and transferred their attention to kangaroos. Hence the commercial kangaroo skin and pet meat industry evolved in the late 1950s and 1960s. At that time kangaroo meat for human consumption, notably in Australia, was frowned upon; but overseas there was an evolving demand, especially in Europe, for novel game meats, and kangaroo meat was identified as fulfilling this niche market.

The current system that allows the commercial harvest of designated kangaroo species follows extensive basic research of their biology, especially their population dynamics, as well as the establishment of rules for their humane killing. Where kangaroo harvests occur, the Australian Government's aim is to ensure that these are sustainable.

Although the overall management of the kangaroo harvest is under the umbrella of the Commonwealth Government, the individual states manage the harvest within their own jurisdictions. Currently the individual states have different procedures in place, but everywhere the industry is highly regulated. Even so, by its very nature much of the harvest is conducted in remote areas, hence adherence to these regulations depends on the integrity of the harvesters.

The ultimate responsibility for the harvesting of kangaroos lies with the Australian Minister for Sustainability, Environment, Water, Population and Communities in the Commonwealth Government. The minister is guided by the *Wildlife Protection (Regulation of Exports and Imports) Act* 1982, which endeavours to ensure that no Australian native species is put at risk through inadequate control of exports or via the import of exotic species. More specifically, the minister is guided by the *Environment Protection and Biodiversity Conservation Act* (EPBC Act), which came into force in January 2002 to approve a wildlife trade management plan for any state or territory for a maximum of 5 years. Responsibility for the sustainable kangaroo harvest is delegated to the individual states and territories, each of which must develop its own management program. This has to be done for each species in question, based on an adequate knowledge of its biology and ecology. Each program must have current Commonwealth approval. The minister has the power to revoke approval if the provisions of the Act have been contravened, and to reinstate approval only when the designated authority demonstrates that the program is properly constructed or that measures have been put in place to remedy any faults.

In this way each state or territory is responsible for ensuring the conservation of harvested species under its care, and for the sustainable harvesting of these designated species. Furthermore, the harvesting processes must adhere to strict health regulations in the production of products, including meat for human consumption and for pet consumption. Inadequate adherence to these regulations in 2009 resulted in cessation of kangaroo meat export to Russia, Australia's major market. This in turn resulted in the loss of more than 200 jobs in the industry. Following an investigation of the industry, the state and Commonwealth regulatory authorities undertook the implementation of better hygiene standards, as well as radio frequency identification tags and barcodes to trace kangaroo product from paddock to export to ensure a resumption of the trade.

The specific powers that states or territories have allow:

- the restriction of the species of kangaroo being harvested
- the limitation of the number of kangaroos being harvested
- stipulation of the size and/or the weight of animals harvested

- stipulation of the method of killing adults and dependent young
- control of the period of harvest
- control of the places of harvest by specifying exactly where animals may or may not be taken
- restriction of the number of permits allocated to commercial shooters in each specified category: human consumption or pet meat
- refusal to issue a new permit
- control of the number and location of dealer sites
- monitoring and assessing of non-commercial take through licence returns.

The *Kangaroo Conservation and Management Plan for South Australia 2008–2012* most clearly articulates the principles that underpin and guide the implementation of the plan, and the achievement of its goals. Similar principles underpin the management of the commercial harvest of kangaroos in all states having this industry.

Table 4.1 Quoting from the plan above, these principles are as follows:

Wildlife is valued	Kangaroos have intrinsic value as species, and are valued for their role in ecosystems and Australian landscapes, and for their ability to be used as a sustainable resource. This value underpins conservation and management efforts.
Ecological capacity	Kangaroos and ecosystems of which they form part must be managed in a manner that is within their ecological capacity and does not pose a risk to their long-term conservation.
Natural resources management	Management of kangaroos is a component of natural resources management, and should be integrated with the management of other natural resources.
Sustainable use	Use of wildlife, when it takes place within ecologically sustainable limits, is an important conservation tool. The social and economic benefits derived from such use provide incentives for people to conserve wildlife (IUCN Sustainable Use Specialist Group 2000).
Ecologically sustainable development	Kangaroo management takes place in accordance with the principles of ecologically sustainable development, including consideration of environmental, economic, social and equitable considerations and the principle of inter-generational equity (EPBC Act).
Multiple outcomes	Management actions for kangaroos should aim to achieve multiple environmental, economic, social and cultural outcomes in an equitable and balanced fashion (in accordance with the ecosystem approach, Convention on Biological Diversity 1995).
Precautionary principle	Where there are threats of serious or irreversible damage or impacts to kangaroos or ecosystems, lack of full scientific certainty should not be used as a reason for postponing measures to prevent or mitigate impacts.
Adaptive management	Kangaroo management is based on an adaptive approach that is flexible and adaptable, designed to adjust to the unexpected, promotes learning through doing and is based on science and knowledge.
Develop and share knowledge	Knowledge provides a platform for management decisions. It is essential to develop and share knowledge, and to seek and value the knowledge of Government, industry, scientists, and rural, urban and Indigenous sectors.
Indigenous values	Indigenous heritage, knowledge and cultural values should be integrated with the conservation and management of kangaroos.
People are the key	Capable, connected and committed people are critical to achieving the sustainable management of kangaroos and natural resources, and should be at the centre of management.
Working together	Government, industry, urban, rural and Indigenous communities must work together with inclusive and transparent decision-making to ensure the effective conservation and management of kangaroos.

Table 4.2 South Australia's options for the kangaroo management plan	
Commercial harvest	A sustainable harvest of kangaroos takes place across the pastoral and agricultural rangelands of South Australia, in accordance with allocated harvest quotas and strict industry regulation.
Non-commercial destruction	Landholders can apply for permits to destroy kangaroos that are causing detrimental impacts or damage. Carcasses cannot enter the commercial trade.
Management of kangaroos on reserves	On conservation reserves, kangaroo populations can increase to levels that are unsustainable for habitats. When this occurs, numbers may need to be managed. This may include culling of kangaroo populations to reduce their impacts on native vegetation and other fauna.
Traditional hunting	Under the *National Parks and Wildlife Act* 1972, Aboriginal people can hunt kangaroos for food and cultural purposes.

Currently in Australia commercial harvesting is primarily of the red kangaroo *Macropus rufus*; eastern grey kangaroo *M. giganteus*; western grey kangaroo *M. fuliginosus*; and euro–common wallaroo *M. robustus*. In 2008, about 2.3 million kangaroos were killed legally across Australia. The states of Queensland and New South Wales account for 85 per cent of Australia's quota, most of which is killed on the extensive rangelands used for sheep grazing, where kangaroo densities can be in excess of 20 individuals per square kilometre. South Australia and Western Australia account for about 14 per cent of the Australian quota. In recent times the red-necked wallaby *M. rufogriseus* and Tasmanian pademelon *Thylogale billardierii*, the two remaining harvest species, have been taken either in small numbers or in some years not at all.

The study of the biology and ecology of each harvested species must include annual monitoring of abundance. This is the basis for setting annual quotas. Quotas for each species vary from year to year because of changing seasonal conditions, notably drought and rainfall. The methodologies and techniques for monitoring abundance vary from simple ground surveys to aerial estimates that are flown along predetermined gridded flight lines. These are statistically derived patterns, and the results are ground-truthed regularly. Additional biennial and triennial aerial transects are also flown to expand the overall picture of the particular kangaroo populations. From the data, scientifically and statistically derived quotas expressed as a 'percentage kill' are determined. Quotas are determined without any reference to industry or market demand. Quotas are usually between 15 and 17 per cent of the estimated population. In 2008 the quota for Queensland was close to 2 million, New South Wales 1.1 million, and the remaining 0.7 million from South and Western Australia combined. Of these, only 58 per cent were harvested, the mismatch being typical of most years. In Tasmania, the 2008 quota for harvesting red-necked wallabies and the Tasmanian pademelons on Flinders and King islands were 55 000 and 15 000 respectively. The harvest of these species from the two islands fluctuates dramatically, according to the changing international demand for product. In recent times there has not been a harvest of these two species.

Broadly speaking, each state having a kangaroo harvesting industry sets down procedures for kangaroo harvester licence applicants to follow. These include record-keeping, tagging, conveyance to a licensed carcass dealer and procurement of licences for export. The documents pertaining to Queensland, New South Wales, South Australia and Western Australia are set out in Appendix C.

Qualifications and attributes essential for a shooter

Under the *Code of Practice for the Humane Shooting of Kangaroos and Wallabies for Commercial Purposes* (COP), a shooter, whether a harvester or not, must be appropriately skilled to assess the terrain and weather conditions to cause a sudden and painless death. Under most conditions a centrefire rifle is used, and all kangaroos are required to be brain-shot in accordance with the 'points of aim' specified in the COP. With most kangaroo shooters and under most field conditions brain-shots are achieved (98 per cent), and the kangaroo instantly dispatched. Following a Commonwealth Government-sponsored audit of kangaroo harvesters, a report *Kangaroo*

Shooting Code Compliance. A Survey of the Extent of Compliance with the Requirements of the Code of Practice for the Humane Shooting of Kangaroos was prepared for Environment Australia by RSPCA Australia in 2002. This documents that in the modern kangaroo harvest 98 per cent of kangaroos taken by licensed kangaroo shooters are dispatched by a high-powered bullet either in the head or at the junction of the skull and neck exactly where the shooter aims.

Under special conditions, such as the dispatch of a wounded animal, a heart shot may be used. Where an animal is injured, the shooter must immediate cease shooting other animals to make every reasonable effort to locate and kill the injured kangaroo humanely. Following their death the pouch of all females must be searched and any pouch young removed. These are dispatched painlessly and as quickly as possible. With small furless pouch young, cervical dislocation is recommended. With all furred pouch young, a single forceful blow to the base of the head killing the young is recommended. With young-at-foot, shooting to destroy the brain or heart is advised. Extensive analysis by the most informed experts in the field has confirmed these are best practice and deliver humane dispatch. However, there still remains the probability that some young-at-foot do escape and die.

In its 2002 report the RSPCA recommended that the use of shotguns be removed from the Code of Practice and that only centrefire rifles be used in wallaby harvesting. Shotguns are not permitted for use in commercial harvesting in states with an approved kangaroo management plan.

Responses to comments for and against the sustainable harvest of kangaroos

1. *Leave the kangaroos to regulate their own population size.*
   In the wild most populations have a dynamic balance of animal numbers determined primarily by the availability of food and water linked secondarily with predation levels. However, most of the Australian landscape has been dramatically affected by the introduction of many novel mammal species, such as sheep, cattle, European red fox and domestic cats. This, coupled with European agricultural management practices, has dramatically altered the pre-European ecological balances. In some cases this has caused a decline in the size and range of many kangaroo species, but in the case of the species currently being harvested, the opposite has occurred.
2. *They are in plague proportions in western New South Wales and Queensland, where they are competing with sheep for graze and are damaging the environment.*
   Under any circumstance where the total grazing capacity (TGP) of the land is exceeded there is serious damage to the environment. This is particularly true during droughts, when sheep and kangaroos compete for the same resources. The commercial harvest is the only tool to manage this component of TGP, and kangaroo management must be integrated with the total management package of the land.
3. *We should use more natural management changes to lower high kangaroo numbers.*
   Changes to grazing management programs may lower kangaroo numbers in some areas. For example, encouraging the growth of perennial grasses while limiting green shoot growth will discourage both red kangaroos and eastern grey kangaroos, as these species have highly specific grazing preferences from which they prefer not to deviate. Domestic livestock can handle rank tall grass but kangaroos avoid this. Closure of dams and the removal of open bore drains have been successful in lowering kangaroo populations in some local locations. However, some animals have had a horrible death from dehydration. Moreover, in many instances these measures simply resulted in the kangaroos relocating themselves to more favourable areas. Given their ability to exist with limited drinking water and to travel long distances to water, this is unlikely to be a viable alternative to ongoing harvesting.
4. *We should use alternate methods to control high kangaroo numbers.*
   Exclusion fencing, especially associated with watering points, is proving to be expensive, but is in some instances a successful method of limiting kangaroo numbers especially for small areas. As most harvesting is in rangeland, exclusion fencing is prohibitively expensive.

Other control methods, such as trapping, deterrents and distractant crops, have been of limited success, but cannot be implemented on a broad scale. However, in all of the above 'solutions' to lower kangaroo numbers the animal welfare outcomes of these are rarely considered. For instance, exclusion fencing may result in slow starvation and dehydration.

5. *We should use contraceptives to control kangaroo numbers.*
   Contraceptive implants and immuno-contraception for kangaroos are prohibitively expensive to use on a broad scale. This is compounded by the need to capture and handle each individual for implantation or injection. Severe stress, injury and even death can happen when capturing and handling wild kangaroos.
6. *A wildlife harvest can never be sustainable, and as such should not be undertaken.*
   The kangaroo harvest has been fully sustainable for more than 30 years. If anything, commercial decisions such as animal densities being too low for harvesting to be economically worthwhile at the time prevent the full takeup of the harvest quotas set across Australia.
7. *Stopping the kangaroo harvest would be advantageous to the sheep farmers.*
   Most of the kangaroo harvest is conducted in the arid rangelands of Queensland and New South Wales. Cessation of the harvest in most of these areas would result in a rapid rise in kangaroo numbers. This in turn would increase the grazing pressure on the environment. The result would be that the pastoralists would be forced to do their own shooting or to employ staff to do so. These people are unlikely to be as well trained and professional as current licensed harvesters. Consequently, animal welfare in particular would decline. Secondly it would be less cost effective than the current system.
8. *Kangaroos are different to domesticated livestock and should not form the basis of a production industry.*
   The world's demand for animal protein for human consumption is high, and is met mostly by meat from sheep, goats, cattle, pigs and chickens. However, in Australia there are large numbers of large grazing kangaroos being killed annually as part of the management of rangeland pastures to lessen the impact of their competition with domestic stock. Use of these animals is making commercial use of products that otherwise would be left in the field.
9. *We should not shoot our national symbol.*
   Kangaroos are highly regarded and valued in the minds of most Australians. Although few people like the idea that any animal is killed to provide meat for their benefit, as long as large numbers of people eat meat, the killing of animals will continue. With the extremely large populations of large grazing kangaroos in Queensland and New South Wales, their sustainable harvest will continue as a co-product of rangeland management.
10. *We should not harvest kangaroos for human use.*
   Broadly speaking, animal rights philosophy contends that animals have basic rights, including the right to life, and thus explicitly rejects the human 'use' of animals for the benefit of humans. This contrasts with the animal welfare perspective, which accepts the use of animals for human consumption, economic benefit, medical testing and so on, if the animals are treated in a humane and justified manner. The ideological difference between animal welfare proponents and animal rights advocates has been described in the context of poultry farming as the difference between 'bigger cages' and 'empty cages' – with 'empty cages' (or the total 'liberation' of animals from human use) being the eventual goal of animal rights supporters. The two positions are quite distinct, and should be recognised as such. In the context of kangaroo harvesting, the 'empty cages' goal is not always apparent. Debate does not centre on the 'liberation' of kangaroos from human use, but rather on the issue of perceived cruelty. Animal rights advocates argue that kangaroo harvesting is 'cruel' and appeal for total bans on the industry. However, the statement that kangaroo harvesting represents animal cruelty is based on the view that killing an animal for human consumption and/or economic benefit is morally wrong, and 'unnecessary' for human survival. In some cases arguments are made for ever-increasing restrictions, representing prohibition in increments.

In short, animal rights advocates hold that the death of an animal at the hands of a human for any reason (excluding euthanasia) is an act of cruelty, irrespective of the way in which the animal is killed or the reasons for killing that animal. From the animal rights perspective, issues such as the economic value of kangaroo harvesting to rural communities and families, or the implications for animal welfare of not controlling kangaroo numbers are disregarded or dismissed as providing no justification for the 'immorality' of the industry. Within the animal rights view, the only 'morally acceptable' course of action is to end the industry, and no codes of practice or changes to the industry will shift this belief.

Trying to reconcile the animal rights view with the economic and social benefits of kangaroo harvesting is like trying to compare apples with oranges, or trying to answer questions of religion by using science – the two remain incompatible. The animal rights view is a belief system based on a particular philosophical position; whereas the animal welfare view is quantifiable, and material to which broader values or dollar values in terms of rural employment and community capital can be attached. Those figures translate into benefits for rural families and communities. Conversely, to end the industry would be a significant blow to those communities in rural Australia that depend on the kangaroo industry for their very survival. However, these practical issues are not accorded legitimacy in the stance of philosophical opposition to kangaroo harvesting adopted by animal rights proponents.

11. *The animal will suffer.*
Under legislation all harvesters must be able to demonstrate current excellence in marksmanship. A head shot results in instant death. Monitoring across all harvest areas shows that 98 per cent of harvested animals are killed by headshots. In addition, unlike in the dispatch of domestic herbivores, there is no stress from mustering, transport or holdover periods in an abattoir. Consequently, with an unexpected headshot there is no possibility of the anticipatory stress associated with the dispatch of domestic herbivores.

12. *When you shoot a female kangaroo in the field, you do not know if it has a pouch young attached to its nipple or if it has a young-at-foot. Consequently you are condemning the young to a slow lingering death.*
Throughout the harvest year there are periods when it is highly likely that most females will have a dependent young. By law, the harvesters have to search the pouch of every female that they kill. Under the National Code of Practice for the Humane Shooting of Kangaroos and Wallabies for Commercial Purposes, and Non-commercial Purposes, if there is a pouch young present, it must be dispatched humanely and quickly. Harvesters are advised to avoid shooting females with young-at-foot. They do so because they do not enjoy dispatching young animals, and also because it is time consuming and decreases their earning capacity. For access to these codes of practice see <http://www.environment.gov.au/biodiversity/trade-use/wild-harvest/kangaroo/pubs/code-of-conduct-commercial.pdf> and <http://www.environment.gov.au/biodiversity/trade-use/wild-harvest/kangaroo/pubs/code-of-conduct-non-commercial.pdf>. Seventy-five per cent of kangaroos harvested are males.

13. *You (government authorities) do not know what is actually happening out 'there' in the field.*
The kangaroo harvest is the most regulated of all wild harvests in the world. There are unannounced inspections (in excess of 1000 in New South Wales alone each year) by government officials at all levels of the industry, with strict enforcement for non-compliance through fines and withdrawal of licence.

14. *I have watched a video clip of barbaric practices in the dispatch of both adult and pouch young kangaroos.*
One such video was made in 1986, and has been shown repeatedly since as 'evidence' of the cruelty of the kangaroo harvesters. The filming was undertaken at the behest of an animal rights group. Overseas film industry personnel filmed an unlicensed shooter who had 'set up' a kangaroo killing scenario designed to shock. The video is abhorrent to all, including those working in the kangaroo harvest industry. The shooter was successfully prosecuted in 1997 in the Dubbo District Court NSW, NP v. Eichner.

15. *It is a threat to the genetic integrity of the kangaroo population.*
    After many sophisticated studies, there is no evidence that there has been any change in genetic diversity among the four kangaroo species that make up 99 per cent of the kangaroo harvest. There is no basis to believe that genetic effects will occur as a result of long-term harvesting.
16. *It could drive the kangaroo population into extinction.*
    There are ongoing annual surveys of the populations of all target species across the entire area of harvest. These use well-developed and proven techniques and technologies to gather the data used to determine harvest quotas. Harvest quotas are conservative, and in most years harvest quotas are not filled. Kangaroos are generally not subject to harvest on Crown land and in many areas of private property. As such, government authorities are extremely responsive to the population dynamics of the target species across their entire range.
17. *Why don't we simply allow a 'kill and leave' system in place in all cases?*
    The sustainable harvest results in a product, not a waste product.

The economics of the kangaroo harvest industry

In 2009, the kangaroo harvest industry was worth about $200 million to Australia. This consisted of about $60 million to harvesters, $80 million for meat and $60 million for skins. Over the past three decades the industry has had its economic ups and downs, but has continued to be a significant revenue generator, particularly in rural and remote Queensland and New South Wales. The industry not only provides many with an income, it also plays a crucial role in limiting kangaroo numbers on a large number of grazing properties. Overall, the industry has proven to be sustainable on both ecological and financial grounds.

The kangaroo harvest industry employs about 4000 people. For many it is their primary source of income, and for others, an important secondary source. Many families can only continue to live in remote areas because they are able to supplement their incomes from kangaroo harvesting. In this way, the industry ensures the long-term viability of some small rural and remote communities. The jobs involved range from the professional harvester, to chiller box operators, truck drivers, abattoir workers, tannery workers, product distributors to sales personnel.

The financial costs to set up as a harvester depend on the type of market that he or she is entering. In all cases a suitable four-wheel drive vehicle, usually a cab type with a trayback equipped with racks, lights and large water containers is needed. When shooting kangaroos for human consumption, the equipment must be of a high standard to ensure the best hygiene possible. Entry costs into the industry vary greatly. At the low end, as in skin-only or pet meat, a typical cost is about $40 000 or more. However, entry into the high end – meat for human consumption – may be as much as $70 000 or more. These costs include the significant level of training and assessment required in order to gain a Harvest licence. Ongoing costs to the shooters, such as fuel, bullets and tags, are high. Generally, the financial return to the harvester is modest, with few making more than $75 000 per annum.

Historically, the greatest numbers of harvested kangaroos were destined for the pet meat trade. However, over the last decade demand for kangaroo meat for human consumption has risen dramatically, and now in excess of 60 per cent of animals harvested are to supply that market. With animals shot for human consumption, extremely strict hygienic field processing and chiller box procedures apply. Once shot and bled, kangaroos are eviscerated carefully and hygienically, leaving their heart, lungs, liver and kidneys in place for veterinary inspection. Animals are tagged prior to transportation, where strict regulations apply to minimise physical and microbial contamination of carcasses. Chillers have strict regulations governing hygiene, and in particular must be able to lower the deep muscle temperature to 7°C within 24 hours. At the kangaroo abattoir following processing, including veterinary inspection, meat cuts are dispersed to a large number of countries around the world. The pet meat is sold mostly on the domestic market.

The harvesting of kangaroo solely for the skin is conducted only in Queensland. Here the skin is removed mostly using a mechanical skinner attached to the tray of the vehicle and the skin tagged appropriately. Even so, this aspect of the commercial harvest is small and declining. Most skins used in commercial production are co-products of the meat trade.

The products

Meat for human consumption

Kangaroo meat has a unique flavour that is popular with the restaurant trade, and increasingly so in many Australian households. A variety of kangaroo meat cuts is available in most major grocery chains, including Coles, Woolworths and Independent Grocers of Australia. It is sold to about 25 countries, with the main markets being the European Union and Russia. It is a fine-textured, soft meat with little connective tissue or fat. When freshly killed and processed, kangaroo meat may resemble beef in flavour; however, as it ages it becomes gamier. More than 50 per cent of the empty body weight of a kangaroo carcass is muscle, and this muscle is concentrated in the loin, rump and thigh. Sheep have a lower carcass yield in weight-for-weight comparisons.

The primary kangaroo cuts are derived from the loin, fillet, rump and shank. These are all ideally suited to pan frying, barbecuing, stir frying and in some cases oven roasting. Because these cuts are very lean, a short cooking time resulting in a medium rare product is best. Secondary cuts are strips and mince, some of which can be used in processed meats. However, currently, other than simple sausages, little commercial processing of kangaroo meat is undertaken.

Kangaroo meat appeals to many people because of its low fat content (1.5–2 per cent) compared to domestic species (Table 4.3). Significantly, kangaroo fat is mostly in the desirable unsaturated form, with both depot and intramuscular fat in kangaroos being less saturated than those of sheep and cattle. It is believed that these differences are a result of the shorter period that food spends in the kangaroo's stomach: about half the time that food remains in the stomach complex of the ruminants.

Interestingly, the level of conjugated linoleic acid, a polyunsaturated fat noted for its anticarcinogenic properties, is extremely high in kangaroo meat – a real bonus for the health conscious.

The protein content (24 per cent) of kangaroo meat is a little higher than that of beef (17–22 per cent), lamb (18–22 per cent), pork (16–19 per cent), chicken (20–23 per cent) and even crocodile (21–22 per cent). Likewise it has relatively high levels of iron and zinc as well as vitamin B_2 (riboflavin), vitamin B_3 (niacin), Vitamin B_6 (pyridoxine) and Vitamin B_{12} (cyanocobalamin).

Meat for pet meat

Roughly 40 per cent of all kangaroos harvested are to supply the pet meat trade, which supports a large number of families. It operates totally independently of the meat for human consumption sector. Although the same stringent animal welfare rules and regulations apply, the field and processing hygiene standards are high but less stringent. Most pet meat is sold on the domestic market.

Leather

Kangaroo leather, a sustainable co-product of the kangaroo meat industry, is a major money earner. The leather is a high-quality tough biological product compared with synthetic-based textiles. However, to produce high quality leather, kangaroo skin must be carefully managed from its source until it is processed.

The internal structure of kangaroo skin differs significantly from that of domestic herbivores, which are the source of most leather. Like all skin, it has an outer thin epidermis (grain layer) and an inner thick dermis (corium). In life, the grain layer is resistant to tearing and is responsible for the animal being waterproof. Significantly, kangaroo skin does not have sweat glands or erector pili muscles, both being structures that weaken the overall skin structure of

Table 4.3 Percentage fat content of meat cuts (approximate)

		Monosaturated	Polyunsaturated	Mono-unsaturated
M. rufus	Red kangaroo	35	32	33
M. giganteus	Eastern grey kangaroo	32	29	39
Ovis aries	Sheep	44	46	10
Bos taurus	Cattle	46	44	10

cattle, sheep and goats. The corium provides strength from dense bundles of large collagen fibres being supported by a mesh of smaller collagen fibres. A high percentage of fibre bundles in the corium are aligned nearly parallel with the outer surface (angle of weave). The angle of weave of kangaroo leather is less than 30 compared to between 45 and 60 in domestic herbivores. This results in kangaroo leather being extremely strong. The skins of grey kangaroos are significantly stronger than those of red kangaroos. The low percentage of fat in the corium also contributes to some of the strength of the leather. Kangaroo leather has 10 times the tensile strength of cow hide.

Historically, leather-processing plants were dirty, smelly, polluting industries. Now modern leather-processing facilities are highly automated efficient enterprises where a large percentage of the water and chemicals used are recycled. There are strict quality control protocols in place at all levels of the tanning process so that environmental pollution is minimised. Converting raw skins from a readily biodegradable tissue into one resistant to degradation requires changing the internal structures of the skin. This is accomplished by tanning, which is the process of altering the proteins within the skin into stable forms that will not decompose. A variety of chemicals, such as tannins, chromium sulphate, formaldehyde and aromatic polymers, achieve this by cross-linking the protein's polymer chains in the skin. In Australia commercial kangaroo skin tanning uses low concentrations, 70–100 ppm, of formaldehyde.

Overall, kangaroo leather is extremely strong, flexible, durable and lightweight. These properties make it in demand for soccer footwear, high-quality dress shoes, leather car seat covers, motorcycle leathers, baseball gloves, golf gloves and even high-quality bullwhips.

The future for sustainable production of kangaroo products

The farming of kangaroos has been an emotive subject since it was first considered and attempted. Although there have always been strong advocates, such as Professor Gordon Grigg of the University of Queensland, for the concept of farming kangaroos, there have always been a large number of sceptics. The bottom line is the economic viability of these enterprises. If the value of the products produced by farming kangaroos is greater than that of alternative enterprises, kangaroo farming will become viable no matter what obstacles there may be. Currently the economic scenarios are not favourable to farming kangaroos. However, one cannot totally reject the long-term possibility of this happening. Since the late 1990s there has been an increased nationwide acceptance of kangaroo meat for human consumption. Simultaneously, there has been an increasing public awareness of the massive environmental degradation that has occurred through much of Australia because of the impact of hoofed animals on soil profiles and vegetation cover. This contrasts dramatically with the lesser effects of the soft-footed kangaroos on their environment. Another factor in favour of kangaroo farming has been the steady decline of sheep numbers. Although in many instances graziers have converted their properties to cattle production, in some areas this is a marginal financial enterprise. When all of these basic issues are considered together, the possibility of farming kangaroos may be getting closer.

That sheep and cattle have become the major meat and milk producers feeding much of the world is no accident. These two species were domesticated early in the history of modern humanity. It is believed that the two subspecies of modern day cattle were domesticated independently, the taurines about 8000 years ago in the Fertile Crescent of the Middle East, and the zebus about 7000 years ago in the Indus Valley of Pakistan. With sheep there were at least three independent domestications occurring about 10 000 years ago in South-west Asia. Long periods of domestication, coupled with long-term genetic selection for desired products, have made these two species tractable, safe and economical for human exploitation. There is a general worldwide acceptance that sheep and cattle are 'good' for any group of people in virtually any environment. Historically there were no environmental impact studies. Now throughout much of the world, meat from sheep, cattle and goat is considered ideal. What a different story it may have been if the early European settlers to Australia had not introduced sheep and cattle and the dietary philosophies that accompanied them. If the early settlers had accepted the possibility of farming the kangaroo, then they may have become a staple part of Australian cuisine rather than being viewed as an exotic form of meat.

What are the possible options for farming?

1. *Wild harvest*

This is the *status quo*. The current kangaroo harvest is the largest wild harvest of a large mammal in the world. The amount and pattern of rainfall is the principal determinant of kangaroo population size. Good seasons with adequate rainfall at the critical times during the growing period provide the best forage for kangaroos, especially the large species. A series of good seasons results in better survival of kangaroos of all age groups and hence numbers that are more likely to sustain harvesting for a longer period, before the costs of the harvest rise when the number of animals declines.

2. *Rangeland*

Strategies encouraging pastoralists to include kangaroos in their economic support base would diversify their income stream as well as encourage the sustainable use of wildlife and promote conservation. Financial gains from diversified enterprises, such as mixed grazing systems incorporating both kangaroos and domestic stock, could provide producers with the incentive and ability to contribute to habitat replacement and protection. If pastoralists could make enough money from kangaroos, they would become an accepted economic part of their enterprises. Some pastoralists have embraced the idea of having tourism on their properties with kangaroos as a critical part of their visitors' experience. This can be a significant additional source of funds.

To date only a few properties have undertaken measures to increase kangaroo production. These measures have involved:

- reducing or eliminating sheep stocking rates, and in some cases cattle, on part or all of a property
- modifying the fencing regimen on the property by removing some internal dividing fences. Some properties have converted to the use of 'kangaroo friendly fencing': types in which a kangaroo is unlikely to become entangled or 'hung up'
- introducing water points to areas lacking a reliable water supply
- altering graze availability such as by the judicious use of fire to encourage a green flush. Fire can also be used to lessen the amount of dry, low nutrient value, tall grasses that most kangaroos find unpalatable
- excluding all herbivores from overgrazed areas and areas where soil erosion is bad.

Even so, few pastoralists know how many kangaroos are actually on their property. They rarely have any idea of seasonal changes in kangaroo numbers and of their utilisation of resources across their properties. The level of interaction with and/or competition of kangaroos with domestic stock is still contentious. Currently there is little knowledge of the specific nutritional requirements of kangaroos and of how best to manage pastures to their benefit. Likewise, there is little knowledge of best husbandry practices for kangaroos. In the near future there will also be the possibility of diversifying the income stream of average pastoral properties with carbon credits from destocking of sheep and cattle. These actions could make kangaroo production more viable. However, in the end, marketing to enhance the financial return from kangaroo products is essential.

3. *Intensive farming*

Currently intensive farming of kangaroos is not practised in Australia. To do so would need participants to adopt and to adapt modern farming practices including: specific husbandry procedures; genetic selection of proven strains or breeds; pasture improvement and appropriate transport arrangements. In addition, to ensure that the value of kangaroo products were to rise sufficiently to enable intensive farming to be a viable pursuit, there would have to be further market research and additional public education about kangaroo leather and meat, specifically how best to cook kangaroo cuts.

All species of kangaroo have a slow growth rate. At 3 years of age an average male western grey kangaroo weighs roughly 35–38 kg and females 22–25 kg. In most jurisdictions the minimum carcass weight at harvest for grey kangaroos and red kangaroos is 16 kg. In contrast, merino sheep at 9 months of age weigh roughly 30–40 kg and cattle at 12 months, between 360 and 400 kg, depending on their age, breed and growing conditions.

These figures confirm that developing intensive farming of kangaroos is not commercially viable at present. However, because kangaroos have a metabolic rate that is roughly 70 per cent that of comparably sized eutherians they are efficient utilisers of rangeland graze. Furthermore, their lower overall need for water means that they are more effective and efficient utilisers of arid–semi-arid zone native pastures. Because kangaroos produce little methane if one factors in carbon dioxide equivalents then the overall equation alters. Although research into carbon dioxide equivalents of meat production from domestic herbivores is still in its infancy, it is accurate enough for us to know that it is expensive. Production of a kilogram of meat from a 20 kg eastern grey kangaroo has an environmental cost equivalent of about 0.4 kg CO_2e/kg hot standard carcass weight. This is considerably lower than the estimated 9.9–12 kg CO_2e/kg hot standard carcass weight for beef or 7–8 kg CO_2e/kg hot standard carcass weight for sheep meat. With the slowly evolving movement towards the establishment of carbon pollution accountability, production of kangaroo meat with a very low carbon footprint will become more and more desirable.

Ultimately, to improve the quality of kangaroo meat and skin it will be necessary to undertake genetic selection and domestication of kangaroos. The common belief is that it takes thousands of years to domesticate a wild species. However, over the past 50 to 60 years scientists in Russian have successfully domesticated the European red fox. In this endeavour, selection for desirable behavioural traits and, in particular, friendliness towards humans was the key to success.

Renowned evolutionary biologist Jared Diamond considered that six characteristics are essential for the successful domestication of a wild animal:

- They should have an agreeable disposition; that is, not dangerous to their fellows or humans.
- They should recognise a dominance hierarchy.
- They should recognise humans as their dominant.
- They must have a flexible diet and be able to live off a wide variety of food sources.
- They must be able to breed readily in captivity.
- They must have a fast growth rate as well as a relatively short lifespan.

If one applies these criteria to the kangaroo species that are potential candidates for domestication one notes the following:

- The disposition of the kangaroo varies, with some species being more disposed to humans than others. The temperament of some red kangaroos can be quite placid.
- Most of the larger species form mobs and have dominance hierarchies.
- They may learn to recognise humans as being dominant.
- They are surprisingly inflexible in their preferred diets. However, they are well adapted to the native pastures of their preferred environments.
- They do breed readily in captivity but are slow.
- They have relatively slow growth rates and have relatively short lifespans.

Once a decision is made to domesticate a species of kangaroo, the animals would be selected and bred for tractability. It could take one to two decades before demonstrably reliable improvements would become measurable. Later in the domestication process specific animal lines would be selected for carcass characteristics, hide quality, fertility and weaning time. As for many new enterprises, management practices more appropriate to the species would evolve concurrently.

The basic factors that need to be taken into account in a kangaroo domestication program include:

- *Legal status of the animals*
  Currently kangaroos and wallabies in all states and territories are the 'property' of the Crown not of the landowner. Consequently, before a landowner can derive a direct income from the macropods on their land there needs to be a transfer of sovereignty from the respective governmental authority to the landowner.
- *Possible species*
  The red kangaroo and the eastern grey kangaroo are major contenders for domestication. Physiologically, red kangaroos are extremely well adapted to the arid, low grass-plain areas, while eastern grey kangaroos are adapted to slightly wetter woodland areas ranging from mallee through to dry sclerophyllous forests with understoreys of grasses, ferns and shrubs. Two other kangaroo species, the western grey and the common wallaroo, are currently harvested, and may also be potential candidates for domestication. However, the western grey

kangaroo can have an unattractive taint to its meat. The euro, especially females, may be too small for domestication because they produce too little meat once dressed.

- *Particular geographic area for farming*
  A basic knowledge of the biology of the kangaroo species in question will determine the areas that are most appropriate to farm them. For example, red kangaroos are well adapted to the extremes of heat in the arid zone, but are less tolerant of cooler wet conditions found in more southerly regions. They would be a poor choice to farm in wet winter rainfall regions.
- *Behaviour*
  Animals need to be sociable and not flighty. They must be amenable to being mustered and yarded. Once contained, they must be approachable by humans. Finally, they will need to be amenable to transportation.
- *Body size*
  They must be large. Current harvesting practices are designed to accommodate animals in the 20-plus kg body weight spectrum. Unless prices rise significantly, smaller animals would be unviable because of their low meat yields and their skins being too small.
- *Meat recovery and quality*
  Kangaroo meat has gained more and more acceptance as an exotic game meat, especially in the international marketplace. To maintain and expand these markets, attention will be needed to ensure a consistency of meat quality, flavour and tenderness. The meat is already known to be low in fat and high in protein, both significant factors in today's diet-conscious world. Selection for a higher percentage of major expensive cuts as well as of a favourable dressed carcass weight is important to the producer.
- *Hide quality*
  Fortunately, kangaroo leather is light, tough and durable. It is also recognised as being an excellent product; however, it needs to be promoted vigorously so that there is a greater awareness of its fine properties, and thus its value as a garment. A greater public awareness of the superb qualities of wearability and warmth of kangaroo fur is needed. Attention to obtaining furs of uniformly high density, length and colour is critical.

Appendix A

Additional guidelines on animal translocations in Australia

1. All decisions, actions and results should be recorded.
2. Selection of the most appropriate available site for receiving the animals is critical.
   - Ecological and behavioural factors must be taken into account following field research on founder animals. Some factors are quite subtle, and crucial to the success or otherwise of a translocation.
   - A site with ready access for all personnel involved in the program increases the possibility of success.
   - Large areas are more difficult to establish, especially if they need to be fenced and/or need intensive predator control.
   - With large areas it may be better to start with a small area within the main block and then move onto the larger area once initial problems have been addressed and solved.
   - If environmental factors are suitable, islands and peninsulas can be excellent sites because predator/competitor control is usually less expensive and easier to attain.
3. A balancing of the pros and cons of whether or not to take action on complicating factors is important. The following questions are pertinent in many translocations:
   - Should competing herbivores, such as rabbits, goats, pigs and camels, be controlled?
   - Is prior environmental rehabilitation, such as revegetation of the reintroduction site, necessary?
4. Determination of the best or most appropriate source of the founder animals is essential.
   - If there are several sources, then the genetically most diverse is preferable.
   - If there is a choice between island and mainland sources, then a mainland source should be used.
   - If the taxonomic status of the potential founder animals is in doubt, genetic research is necessary to confirm their suitability as a founder population.
5. Evaluate the potential parasitic and infectious disease status of the founder population. If this is positive, then determine the possible risk factors within any potentially newly introduced population as well as determine risk factors to sympatric species at any new introduction site.
6. Where possible use a mid-sized founder group in the initial translocation.
   - If possible, this means 40–50 animals. This allows natural aggregations, and in particular appears to lessen the demands for an individual's vigilance for predators. Where possible it is preferable to use a mix of unrelated individuals rather than small family groups. This ensures that the greatest genetic mix possible is achieved.
   - If fewer animals are available (say less than 10) several top-up translocations will most probably be needed.
7. Identify and minimise stress factors to the founder animals involved. These include climatic factors, capture, handling, transport and final release. A full veterinary clinical examination of all founder animals is essential. This includes a full physical examination, faecal examination, haematology and serum biochemistry, serology and microbial culture. The individual animal's ongoing welfare is a critical factor in all handling and manipulations.
8. Determine the most appropriate method of release for the species in question under the particular circumstances. Consideration of the conditions for animal release might include:
   - As soon as possible into their new location.
   - Holding them in temporary enclosures for initial acclimation over several days to weeks.
   - Holding them long term *in situ* in custom-built breeding pens and then harvesting

animals from there to introduce into their new location.

- Are the animals to be released directly into the environment without supplementation of food and/or water (known as 'hard release')?
- Are the animals to be released into the environment with supplementary food and/or water (known as 'soft release')?

9. Determine suitable monitoring protocols.
   - Institute appropriate methodologies.
   - Ensure the correct monitoring equipment is available.
   - Ensure monitoring will continue for the lifetime of the program; this could be for several decades or longer.
10. Establish the criteria for success or otherwise of the translocation program. For example, the Western Australian Department of Environment and Conservation uses the following criteria:
    - initial mortality of founders (less than 30 per cent)
    - body condition of animals equal to or better than that of the founder population
    - breeding success
    - recruitment to the population
    - first generation breeding
    - population expansion
    - population dispersal
    - nth generation breeding.
11. During and after any translocation any dead animals must have as complete a post mortem as possible. This should include histopathology and microbial culture.
12. Determine whether or not human uses of the translocation area can occur in tandem with the animal translocations.

Appendix B

Recovery Plan statutory provisions and objectives

The statutory provision of Recovery Plans for nationally listed threatened species, as required by the Commonwealth *Environment Protection and Biodiversity Conservation (EPBC) Act* 1999

The Commonwealth *Environment Protection and Biodiversity Conservation Act* 1999 requires that a Recovery Plan be drawn up for any nationally listed threatened species. This will describe the main threats to a species and describes what actions are necessary to counter those threats. In developing a Recovery Plan the authors must liaise with other interested parties, such as state/territory agencies, individuals and groups affected by or particularly interested in the plan, including appropriate Indigenous peoples, as well as ensure the plan complies with Australia's international obligations. Once a draft plan is completed it is submitted to the Australian Government Department of Sustainability, Environment, Waters, Population and Communities (DSEW&C) or to the relevant state/territory agency. The plan is then opened for public consultation, after which it can be redrafted then submitted to DSEW&C, who, following consultation with appropriate authorities, advise the Minister on whether or not to adopt the plan. If adopted, a public notification is posted.

A basic Recovery Plan identifies the species in question; its national distribution; the area covered by the plan; and a description of actual and potentially suitable habitats. Habitats critical to the long-term survival of the species must be identified except under circumstances where if that were done the species would be even more at risk, such as from collectors. Aspects of the animal's biology that impact on the likelihood of survival are also important. Likewise, the identification of all threats – predators, disease, and habitat risks, such as drought and fire – are all crucial. Once the species information, distribution and location, and threats have been determined, the nuts and bolts of the plan can be developed. This entails setting out clear objectives, performance criteria and actions.

2002 Commonwealth Recovery Plan Guidelines on Objectives

The Commonwealth's Recovery Plan Guidelines of 2002 note the following: 'Objectives may address (but not be limited to) issues such as: improved conservation status; reduction in rate of decline; abatement of threats; and increases in number or areas of occurrence, areas of habitats and overall distribution.' It is important that if possible the plan's objectives should be measured quantifiably. Ultimately the plan needs to identify which individuals or organisations will undertake the evaluation of the plan. Regular, annual evaluations are important to enable ongoing adjustments to the plan's yearly activities. The final aspect of a plan is to set out all the actions necessary to achieve the plan's objectives, identify which organisations are to do the work, and determine the financial implications for all involved parties.

Appendix C

Licence application procedures for harvesting kangaroos from Queensland, New South Wales, South Australia and Western Australia

Queensland

In Queensland all native wildlife is protected under the *Nature Conservation Act* 1992. For harvesting to occur, a species must be classified as being 'of least concern' according to the *Nature Conservation (Wildlife) Regulation* 2006. Harvesting can then be approved under the *Wildlife Trade Management Plan for Export – Commercially Harvested Macropods: 2008–2012*.
The Queensland Environmental Protection Agency (EPA) administers the commercial kangaroo harvest throughout the state. A successful applicant for a harvester's licence must have a valid and current Firearms Licence that allows the person to possess and use an appropriate firearm. The person must be fully firearm accredited under the *Weapons Act* 1990.
In all field operations the harvester must comply with the current *Code of Practice for the Humane Shooting of Kangaroos and wallabies for Commercial Purposes*.
In addition to this, purchasers of kangaroo skins or carcasses need to have a Commercial Wildlife Licence for dead macropods under the *Nature Conservation (Administration) Regulation* 2006.

Applicant for Commercial Wildlife Harvesting Licence
Obtains Harvesters licence

↓

- all criteria met and licence issued to harvest macropods,
- licence specifies the properties on which the harvester can harvest,
- approval of property owner beside each property listed on application,
- tags and 'record and return of operations' book issued,
- harvester shoots kangaroos,
- harvester tags kangaroos before leaving property; places particulars in 'record and return' book,
- harvester delivers macropod carcasses or skins to licenced commercial wildlife licence for dead macropods holder (dealer),
- provides return of operations to Environmental Protection Agency (EPA) within 14 days of the end of each month.

Applicant applies for Commercial Wildlife Licence
(Dealer's licence)

↓

- Dealer's licence can be issued to an individual or a company,
- all criteria met, licence issued to buy dead macropods,
- licence specifies the full address of the licensed site where the dead macropods can be purchased,
- record and return of operations book issued,
- location of where record books will be kept in a secure manner,
- dealer buys dead macropod carcasses or skins from harvester,
- dealer provides harvester with receipt for purchase,
- dealer completes movement advice and forwards carcasses or skins to processors,
- dealer provides return of operations to EPA within 7 days of the end of each month,
- whole carcasses, skins and processed meat can be exported from Queensland.

Procedures covering the commercial shooting of kangaroos and the processing of carcasses and skins in Queensland.
Source: *The Queensland Wildlife Trade Management Plan for Export – Commercially Harvested Macropods 2008–2012.*

Queensland is a very large state with quite different terrains, climates and ecosystems resulting in disparate biological situations in distinctly different areas. Consequently harvest quotas are developed separately for eastern, central and western geographic zones. The majority of Queensland's harvest is from the central zone and then the eastern zone followed by the western zone. Most coastal regions and Cape York are non-harvest areas. Additionally, gazetted parks under the *Nature Conservation Act* 1992 and tenured forests under the *Forestry Act* 1959 are non-harvest areas.

New South Wales

The following information comes directly from the *Commercial Kangaroo Harvest Management Plan 2007–2011* for New South Wales, Sydney.

In New South Wales all kangaroo species are protected fauna under the *National Parks and Wildlife Act* 1974. The New South Wales Department of Environment, Climate Change and Water is responsible for the 'protection and care of fauna'. However, provision for the harvest of kangaroos in New South Wales is regulated under the *National Parks and Wildlife Act* 1974 and by the New South Wales *National Parks and Wildlife Regulations* 2002.

Under section 121 of the Act, a landowner can apply for an Occupier's Licence to allow commercial harvesting on his or her land. A successful applicant is allocated a specific number of numbered tags, one for each authorised animal killed under clauses 48 and 49 of the *National Parks and Wildlife Regulations* 2002. The landowner can nominate authorised harvesters to shoot kangaroos on his or her property. All such persons must hold a Commercial Fauna Harvesters (trapper) Licence under section 123 of the Act. All such persons must also have successfully completed the New South Wales TAFE course 5725 – Australian Game Meat Hygiene and Handling within 3 months of being licenced, and have a current Firearms Licence.

Sale of kangaroos is solely to Fauna Dealers, Licenced according to the NPW Act 1974 section 124.

Persons holding a Skin Dealer's Licence under section 125 of the Act can only undertake the purchase of skins. The import and export of kangaroos into or out of NSW is by an Import and Export Licence under section 126 of the NPW Act 1974 and clauses 50 and 51 of the *National Parks and Wildlife Regulations* 2002.

Landholder applies to DEC for an occupier's licence to shoot kangaroos

↓

Landholder nominates licenced trapper to shoot commercially

↓

Occupier's licence assessed and issued by an authorised DEC officer.
Licence specifies the number of each kangaroo species authorised to be shot.
The number authorised to be shot must be within the quota.
Plastic tags issued with license (NPW Regulation).
Trapper shoots kangaroos.
Trapper attaches tags to kangaroo carcasses.
Trapper delivers kangaroo carcasses to premises registered by fauna dealer.
Fauna dealer transports kangaroos to registered premises for processing or sells to another fauna dealer.
Skin removed and meat and skin processed.
Kangaroo skin processed at registered premises.
Tag removed during processing of skin.
Whole carcasses and skins, processed meat and processed skins can be exported from New South Wales or imported into NSW pursuant to licence issued under s126 of the NPW Act.

Procedures covering the commercial shooting of kangaroos and the processing of carcasses and skins in New South Wales.
Source: *New South Wales Kangaroo Management Plan 2007–2011*.

South Australia

The South Australia, Department of Environment and Natural Resources (DENR) authorises the commercial harvest of red kangaroo, western grey kangaroo and euro under the *National Parks and Wildlife Act* 1972. The *Kangaroo Conservation and Management Plan for South Australia 2008–2012* is the approved management plan to facilitate this.

Landholder chooses to participate in commercial harvest

Commercial harvest quota is obtained

- Property must fall within a region that has available commercial harvest quota for that given calendar year (outside of the commercial harvest zone, landholder may apply for a non-commercial destruction permit).
- If the regional commercial harvest quota has been fully issued but land management issues remain, Special Land Management Quota may be available.
- Landholder must authorise the use of quota on their property (currently quota is allocated to individual properties; within the life of the plan this may change amending quota issue to zone-based allocation).
- Landholder must provide written permission allowing Kangaroo Field Processor(s) to harvest on their property.

Industry obtains sealed tags for harvest

Sealed tags must be purchased from DENR prior to harvest taking place

- Industry purchases sealed tags in accordance with regulations (currently the Meat Processor purchases the sealed tags from DENR; within the life of this plan, this system is likely to change allowing for Field Processors to purchase sealed tags).
- Tags are purchased for each species that is to be harvested (currently there is a separate colour tag for each species; within the life of this plan, this system may change to having one colour of tag for all three species).
- Tags are nominated to a specific area of use (e.g. currently a property) and to a specific Field Processor (the specifics of this system will be likely to change within the life of this plan to reflect changes to tag sale and quota issue).

Kangaroos are harvested by Field Processor

Field Processor harvests kangaroos in accordance with available quota

- Field Processor liaises with landholder regarding harvesting of kangaroos.
- Each harvested kangaroo has a sealed tag affixed to it after harvest.
- All kangaroos are taken in accordance with the Code of Practice for the Humane Shooting of Kangaroos and Wallabies for Commercial Purposes (within the life of this plan regulations that prohibit the sale of body-shot carcasses will be introduced).
- Field-dressing of the carcass is conducted. Carcasses can only be sold to Meat Processors in approved forms for sale, as defined by regulations.
- Carcasses are temporarily stored in a field chiller, before being transported to a processing works.

Carcasses and skins are processed by Meat Processor and/or Skin Tanner

Kangaroo carcasses and skins are processed

- Sealed tags are removed from skins during the process.
- Carcasses and skins imported from interstate for processing in SA must have been taken in accordance with an accredited interstate kangaroo harvest program, and their import requires an import permit issued by DENR.

Kangaroo meat and skin products are sold to national and international markets

Products are sold to national and international markets

- Export of products from SA requires an export permit issued by DENR.
- Export of products from Australia requires an export permit issued by DEW.

Procedures covering the commercial shooting of kangaroos and the processing of carcasses and skins in South Australia.
Source: *The Kangaroo Conservation and Management Plan for South Australia 2008–2012*, Appendix 2.

Under the commercial harvest, quotas are set for each of the three species being harvested. Quotas are allocated to four harvest regions: Western Pastoral, Eastern Pastoral, Western Agricultural, and Eastern Agricultural. Quotas are based on a sound scientific knowledge and ongoing surveys of the species in the field. In recent times only about 50 per cent of the quotas for reds and western greys have been filled. This figure is lower for the euro, where less than 15 per cent of the quota has been taken. Since 1997 the harvest has fluctuated greatly, and only between 30 and 50 per cent of the quota has been taken.

Western Australia

The following summary of the Western Australian scene comes from the *Management Plan for the Commercial Harvest of Kangaroos in Western Australia 2008–2012*.

All native fauna in Western Australia is protected by the *Wildlife Conservation Act* 1950. Areas such as national parks, nature reserves, State Forest and timber reserves are managed by the Western Australian Department of Environment and Conservation (DEC) under the Western Australian *Conservation and Land Management Act* 1984. Generally commercial harvest is not permitted on these lands. However, if necessary, DEC manages the non-commercial culling of kangaroos on land under their control as well as facilitating damage mitigation schemes.

Outside of DEC-managed areas, commercial harvest of kangaroos is regulated under the *Wildlife Conservation Regulations* 1970 (Regulation 6). This is facilitated by the *Agriculture and Related Resources Protection Act* 1976 (ARRP Act), where the red and western grey kangaroos are listed as declared animals throughout Western Australia. This declaration requires the development of a management program outlining areas and conditions under which controls may be applied. Harvest licences are linked to 22 specific harvest zones. These lie west of a line drawn from just north of Port Hedland to Eucla on the SA/WA border.

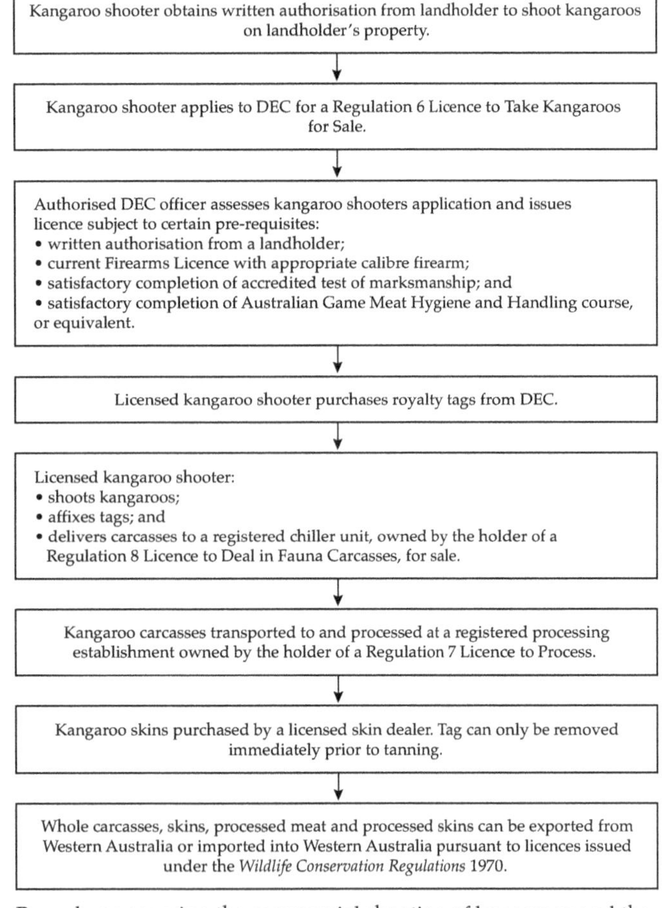

Procedures covering the commercial shooting of kangaroos and the processing of carcasses and skins in Western Australia.
Source: *Management Plan for the Commercial Harvest of Kangaroos in Western Australia 2008–2012.*

Glossary

Allele: an alternative form of a gene (one member of a pair) that occupies a specific position on a specific chromosome. If paired alleles are the same, the organism is said to be homozygous for that trait; if they are different, the organism is heterozygous. These DNA codings determine distinct traits that can be passed on from parents to offspring.

Allelic diversity: a measure of genetic diversity within a population.

Allopatry: the occurrence of populations or species in separate geographical areas with no overlap.

Anaerobic: the absence of oxygen.

Arboreal: living in trees.

Arid: a region with an annual rainfall less than 250 mm. In Australia much of the arid zone is hot desert.

Auditory bulla: a hollow bony structure, part of the middle ear, on the ventral, posterior portion of the skull.

Biodiversity: the variability among all living organisms at all levels within biological systems.

Bipedal: having two feet.

Blastocyst: an embryo at a very early stage of development, consisting of a hollow ball or cells with no differentiation of organs.

Bottleneck: a sudden restriction in population size. This reduces the genetic variation in the population and the population's ability to adapt to new selective pressures.

Browsing: a type of herbivory where the herbivore feeds on leaves, twigs, shoots, flowers and fruit of shrubs and trees.

Catastrophe: an extreme environmental fluctuation that has an extreme negative effect on a population. This could be a novel disease, cyclonic event or extreme bushfire affecting a discrete population.

Cellulose: is the structural component of the primary cell wall of green plants.

Cenozoic: an era in the geological time scale, from about 65 million years ago until the present day.

Chenopod shrubland: salt-tolerant shrubs from the family Chenopodiaceae, covering about 8 per cent of Australia's arid zone.

Chromosome: a single coiled piece of DNA within the cell nucleus that transmits genetic information.

Cloaca: the terminal segment of the gut that also receives the ducts from the kidneys and reproductive organs.

Colony: an association of individuals of the same species.

Comminute: to break down.

Community: an association of populations of different species living together in a defined habitat with some degree of interdependence.

Conservation status: the likelihood that a species may become extinct over a given period of time.

Corpus luteum: the temporary endocrine tissue in the ovary that develops from ruptured follicles after ovulation; secretes progesterone and oestrogen.

Crepuscular: active at twilight or just before dawn.

Critically endangered: a species that has a very high probability of extinction within a short timeframe.

Crus: pertaining to the leg.

Cryptic: an animal's colours and patterns that blend in with the physical surroundings and help it hide from predators.

Cull: to kill animals as a means of population control.

Delayed implantation: where the embryo does not implant immediately in the uterus, but remains in a state of dormancy until it implants at a later date.

Deoxyribonucleic acid (DNA): the long, double-stranded, helical molecule in which the arrangement of nucleotide bases (adenine, cytosine, thymine and guanine) determines the inherited traits of an organism.

Diapause: a temporary halt in the growth and development of an embryonic mammal.

Dicotyledon: a flowering plant having two cotyledons in the seed and leaves with a network of veins radiating from a central main vein.

Diurnal: active during the day.

Dominance: a relationship between two variant forms (alleles) of a single gene, in which one allele masks the expression of the other in influencing a trait.

Ecosystem: a community of living organisms and the physical environment with which they interact.

Emigration: the movement of individuals out of a population.

Endangered: a species or population that has a high probability of extinction within a short timeframe.

Environment: the set of external influences acting on an organism.

Epaxial: dorsal to the axis; often refers to structures lying above the transverse axis of the vertebral column.

Epoch: a subdivision of a period of geological time during the Cenozoic era.

Era: a major subdivision of the geological time scale, marked by abrupt changes in the fossil record.

Eutherian: a major division of placental mammals that have characteristic features of their feet, ankles, jaws and teeth, distinct from marsupials.

Evolution: genetic change in the characteristics of a population of organisms over generations.

Extant: still in existence.

Extinction: the elimination of a species; that is, when the last member of a species dies.

Feral: a domesticated plant or animal that has run wild.

Fixation: where all individuals of a population have the same allele at a specific locus.

Forestomach: nonglandular region of the stomach in front of the true glandular region of the stomach.

Founder effect: the reduction in genetic variation and the change in allele frequencies that occur when a small number of individuals from a larger population founds a new population.

Frugivore: a fruit eater.

Gene: a segment of DNA that serves as a unit of hereditary information.

Gene flow: the exchange of genes among populations by migration and interbreeding.

Genetic diversity: refers to the total number of genetic characteristics in the genetic makeup of a species. It is distinguished from genetic variability, which describes the tendency of genetic characteristics to vary.

Genetic drift: a random change in allele frequency in a small breeding population.

Genetic selection: selection of animals as breeding stock based on known inherited characteristics.

Genetic variance: proportion of the phenotypic variance due to genetic differences among individuals.

Genetic variation: a phenotypic variance of a trait in a population attributed to genetic heterogeneity.

Genotype: the genetic constitution of an organism.

Gestation: the period of development of young in viviparous animals from fertilisation to birth.

Grazing: a type of herbivory where the herbivore feeds on herbage such as grasses or algae.

Haustrum: a pouch of the alimentary tract. Generally found in the stomach or large intestine.

Heathland: an area of nutrient-poor soils dominated by low shrubs.

Heritability: the proportion of the variation for a quantitative character due to genetic causes.

Herbivore: an animal that feeds on plants.

Heterozygous: having a pair of unlike alleles for a specific locus.

Home range: a geographic area that an individual animal seldom or never leaves.

Homozygous: having a pair of identical alleles at a specific locus.

Hybrid: the offspring of two genetically dissimilar parents.

Hybridisation: interbreeding between members of two different taxa.

Hydrology: the study of the movement, distribution and quality of water.

Hypaxial: ventral to the axis; often refers to structures lying below the transverse axis of the vertebral column.

Inbreeding: the mating of genetically similar individuals; homozygosity increases with each successive generation of inbreeding.

Inbreeding coefficient: the probability that the two genes present at a locus in that individual are identical by descent. This ranges from 0 to 1.

Inbreeding depression: depression of performance – for example, low fertility rate and high juvenile mortality – due to inbreeding.

Immigration: the movement of individuals into a population.

Introduced species: a plant or animal species that has become established in an area where it is not native.

Introduction: moving wild-collected or captive-bred animals to areas outside of their historical range.

Invasive species: a foreign species that, when introduced into an area where it is not native, upsets the balance among the organisms living there and causes economic or environmental harm.

IUCN: the abbreviation for the International Union for Conservation of Nature.

Lactation: the production of milk from the mammary glands.

Lacrymal bone: the smallest bone of the face lying in front of the medial wall of the orbit.

Lignin: a complex chemical compound that is an integral part of the secondary cell walls of plants. It is in large quantities in wood and straw and is indigestible to vertebrates.

Locus: a segment of DNA on a chromosome.

Macropores: cavities in the soil that are larger than 75 μm created by agents such as plant roots, soil cracks or soil fauna. Macropores allow water to infiltrate faster or for shallow ground water to flow faster.

Marsupial: a mammal of the infraclass Metatheria, which produces viviparous young that are born in an undeveloped state and have a proportionately long period development before becoming independent of the mother.

Masticate: to chew.

Masseter: a large laterally placed chewing muscle.

Mesic: a habitat with a moderate supply of moisture.

Mesozoic era: the geological time interval spanning 250 million years ago to 65 million years ago.

Metabolic rate: energy used by an organism per unit time.

Metabolism: the sum of all physical and chemical processes that occur within a cell or organism, the transformation by which energy and matter are made available for use by the organism.

Metapopulation: a group of partially isolated populations of the same species that undergo extinctions and recolonisations.

Monocotyledon: a flowering plant having one cotyledon in the seed, leaves with roughly parallel veins and flower parts in multiples of three.

Mutation: a change in DNA producing different effects on an animal's phenotype.

Mybp: million years before present.

Nocturnal: being active during the night.

Occlusion: when the jaws are closed bring the upper and lower teeth into contact.

Oestrus cycle: the recurring physiologic changes that are induced by reproductive hormones in mammalian placental mammals.

Omnivore: species that eat both plant and animals as its primary food sources.

Palaeontology: the study of prehistoric life.

Paracloacal glands: glands lying beside the cloaca. Their principal function is the production of pheromones.

Pectoral girdle: the set of bones that connect the forelimb to the axial skeleton on each side.

Pelage: the coat of an animal consisting of fur, hair wool or soft covering distinct from skin.

Pelvic girdle: complex of bones that connects the trunk and hindlimbs.

Perennial: plants that generally live longer than 2 years.

Pes: foot.

Phalangerid: a member of the marsupial family Phalangeridae, the nocturnal possums and cuscuses of Australia and New Guinea.

Phenotype: the physical or chemical expression of an organism's genes.

Phenotypic variance: variance of the phenotype due to genotypic and environmental factors combined.

Pheromone: a secreted or excreted chemical that results in a behavioural response by other members of the species.

Philopatry: the behaviour of remaining within, or returning to, an individual's birthplace.

Plantar: caudal aspect of the distal hind leg.

Plantigrade: walking with the digits and metatarsals/ metacarpals flat on the ground.

Phylogenetic: pertaining to the evolutionary relationships between groups of animals or plants.

Pleistocene period: the epoch from 2 588 000 to 11 700 years before the present.

Polygamy: a mating system in which individuals mate with more than one of the opposite sex.

Polygynous: a mating system in which the males mate with several females.

Population: a group of organisms of the same species that live in a defined geographical area at the same time.

Precocial: species where the young are relatively mature and mobile from the moment of birth, requiring little parental care.

Quadritubercular: having four tubercles/cusps.

Quadruped: an animal having four feet.

Radiation: evolution of several new species from an ancestral species.

Range: the area where a particular species occurs.

Recovery Plan: the collection information and management documents that sets out actions to assist threatened taxa and ecological communities within a planned and logical framework.

Red List of Threatened Species: an international list from the IUCN of species classified according to perceived risk of extinction.

Reintroduction: the release of individuals of a species into an environment where they have become extinct.

Refugium: the location of an isolated population of a once widespread animal.

Rhinarium: the naked surface around the nostrils of the nose.

Riparian: the habitat for plants and animals on the banks of rivers or other water bodies.

Salinisation: the accumulation of soluble salts in the landscape. Classified as primary (as seen in salt lakes and salt deserts) or as secondary (a result of human activity such as seen following clearing of deep-rooted native plants or following irrigation of crops in areas that have saline ground water).

Salinity: the total amount of dissolved ions in water, dominated by sodium and chloride in lakes and estuaries.

Saltation: to leap as occurs in hopping.

Sclerophyllous: woody plants that have hard leaves and short internodes.

Selection: the change in allele frequencies from one generation to the next in a population as a result of the differential reproductive success of different genotypes.

Semi-arid: a region having an unpredictable annual rainfall of 250–500 mm. In Australia this surrounds the arid desert regions.

Sexual dimorphism: marked phenotypic differences between two sexes of the same species.

Sex-linked: determined by a gene located on a sex chromosome.

Species: one or more populations whose members are capable of interbreeding in nature producing fertile offspring and do not breed with other members of other species.

Squat: site where potoroids either rest up and hide or build their globular nest.

Sternal glands: scent glands in the upper middle of the chest.

Sympatric: populations that share the same or overlapping geographical distributions.

Syndactyly: where the second and third toes of the hindlimb are held close together from their base until distally where the claws are separated.

Taxon: a taxonomic unit.

Tectonic plate: a large irregularly shaped slab of solid rock that makes up the foundation of the earth's crust. There are ten major plates and many smaller ones. Plate size varies greatly.

Threatened: a population or species that has a finite risk within a relatively short timeframe. The IUCN schema considers all critically endangered, endangered and vulnerable categories as being threatened.

Toxoplasmosis: infection due to the protozoan parasite *Toxoplasma gondii*. The primary host is the cat. Morbidity and mortality in some macropodid species can be high.

Translocation: the movement of an individual from one location to another as a result of human actions.

Truffle: the underground fruiting body of a fungus. These fungi have a close association with the roots of vascular plants.

Vulnerable: a species or population with a tangible risk of extinction within a moderate time.

Waterer: an automatic watering device that enables animals to drink the necessary amount of water at any time of the day or night.

Xeric: dry conditions; an environment containing little moisture.

Bibliography

Archer M, Flannery T and Grigg G (1985) *The Kangaroo*. Weldon, Sydney.

Archer M, Hand SJ and Godthelp H (1991) *Riversleigh: The Story of Animals in Ancient Rainforest of Inland Australia*. Reed Books, Australia.

Armati PJ, Dickman CR and Hume ID (Eds) (2006) *Marsupials*. Cambridge University Press, Cambridge.

Breeden S and Breeden K (1966) *The Life of the Kangaroo*. Angus & Robertson, Sydney.

Calaby JH and Flannery TF (2005) *Marsupials of Australia. Volume 3: Kangaroos, Wallabies and Rat-kangaroos*. Mallon Publishing, Melbourne.

Cary G, Lindenmayer D and Dovers D (Eds) (2003) *Australia Burning: Fire Ecology, Policy and Management Issues*. CSIRO Publishing, Melbourne.

Claridge A, Seebeck J and Rose R (2007) *Bettongs, Potoroos and the Musky Rat-kangaroo*. CSIRO Publishing, Melbourne.

Clayton M, Wombey JC, Mason IJ, Chesser RT and Wells A (2006) *CSIRO List of Australian Vertebrates: A Reference with Conservation Status*. CSIRO Publishing, Melbourne.

Coulson G and Eldridge M (2010) *Macropods: The Biology of Kangaroos, Wallabies and Rat-kangaroos*. CSIRO Publishing, Melbourne.

Croft DB and Ganslosser U (Eds) (1996) *Comparison of Marsupial and Placental Behaviour*. Filander Verlag, Furth.

Cronin L (1991) *Key Guide to Australian Mammals*. Reed Books, Australia.

Curtis LK (2005) *Green Guide to Kangaroos and Wallabies of Australia*. New Holland Publishers, Sydney.

Dawson TJ (1995) *Kangaroos: Biology of the Largest Marsupials*. University of New South Wales Press, Sydney.

Diamond J (1997) *Guns, Germs, and Steel: The Fates of Human Societies*. W.W. Norton, New York.

Dickman C and Woodford Ganf R (2007) *A Fragile Balance: The Extraordinary Story of Australian Marsupials*. The University of Chicago Press, Chicago.

Domico T and Newman M (1993) *Kangaroos: The Marvellous Mob*. Facts on File, New York.

Egerton L and Lochman J (2009) *Wildlife of Australia*. Allen & Unwin, Sydney.

Flannery TF (1995) *Mammals of New Guinea*. Australian Museum/Reed Books, Australia.

Flannery TF (2004) *Chasing Kangaroos*. Grove Press, New York.

Flannery TF, Martin R and Szalay A (1996) *Tree Kangaroos: A Curious Natural History*. Reed Books, Australia.

Frankham R, Ballou JD and Briscoe DA (2002) *Introduction to Conservation Genetics*. Cambridge University Press, Cambridge.

Frith HJ and Calaby JH (1969) *Kangaroos*. FW Cheshire, Melbourne.

Grigg GC, Jarman P and Hume ID (Eds) (1989) *Kangaroos, Wallabies and Rat-kangaroos*. 2 volumes. Surrey Beatty & Sons, Sydney.

Hume ID (1982) *Digestive Physiology and Nutrition of Marsupials*. Cambridge University Press, Cambridge.

Hume ID (1999) *Marsupial Nutrition*. Cambridge University Press, Melbourne.

IUCN (2011) *IUCN Red List of Threatened Species. Version 2011.1*. IUCN, Gland.

Jackson SM (2003) *Australian Mammals: Biology and Captive Management*. CSIRO Publishing, Melbourne.

Jackson S and Vernes K (2010) *Kangaroo: Portrait of an Extraordinary Marsupial*. Allen & Unwin, Sydney.

Johnson C (2006) *Australia's Mammal Extinctions: A 50 000 Year History*. Cambridge University Press, Cambridge.

Johnson P (2003) *Kangaroos of Queensland*. Queensland Museum, Brisbane.

Martin R (2005) *Tree-kangaroos of Australia and New Guinea*. CSIRO Publishing, Melbourne.

Maxwell S, Burbidge AA and Morris K (Eds) (1996) *The 1996 Action Plan for Australian Marsupials and Monotremes.* Wildlife Australia, Canberra.

McCullough DR and McCullough Y (2000) *Kangaroos in Outback Australia: Comparative Ecology and Behaviour of Three Coexisting Species.* Columbia University Press, New York.

Menkhorst P and Knight F (2011) *A Field Guide to the Mammals of Australia.* 3rd edn. Oxford University Press, Melbourne.

Ride WDL (1970) *A Guide to the Native Mammals of Australia.* Oxford University Press, Melbourne.

Roache M (2011) *The Action Plan for Threatened Australian Macropods 2011–2021.* World Wide Fund for Nature, Sydney.

Triggs B (2004) *Tracks, Scats and Other Traces: A Field Guide to Australian Mammals.* Revised edn. Oxford University Press, Melbourne.

Troughton E (1967) *Furred Animals of Australia.* 9th edn. Angus & Robertson, Sydney.

Turner JR (2004) *Mammals of Australia: An Introduction to Their Classification, Biology and Distribution.* Pensoft, Sofia, Bulgaria.

Tyndale-Biscoe H (1973) *Life of Marsupials.* Edward Arnold, London.

Tyndale-Biscoe H (2005) *Life of Marsupials.* CSIRO Publishing, Melbourne.

Tyndale-Biscoe H and Renfree M (1987) *Reproductive Physiology of Marsupials.* Cambridge University Press, Cambridge.

Van Dyke S and Strahan R (Eds) (2007) *The Mammals of Australia.* 3rd edn. Reed New Holland, Sydney.

Watts D (1998) *Kangaroos and Wallabies of Australia.* New Holland Publishers, Sydney.

Watts D (2002) *Tasmania Mammals: A Field Guide.* Peregrine Press, Tasmania.

Wilson M and Croft DB (Eds) (2005) *Kangaroos: Myths and Realities.* 3rd edn. The Australian Wildlife Protection Council, Melbourne.

Younger RM (1988) *Kangaroo: Images through the Ages.* Century Hutchinson, Melbourne.

WEBSITES

Animals Australia. <http://www.animalsaustralia.org/>

Australasian Wildlife Management Society. <http://www.awms.org.nz/>

Australian Conservation Society. <http://www.acfonline.org.au/Default.asp?c=2039>

Australian Government, Department of Sustainability, Environment, Water, Population and Communities. Biodiversity. <http://www.ea.gov.au/biodiversity/index.html>

Australian Government, Department of Sustainability, Environment, Water, Population and Communities. Databases and maps. <http://www.environment.gov.au/erin/index.html>

Australian Government, Department of Sustainability, Environment, Water, Population and Communities. Species profile and threats database. <http://www.environment.gov.au/cgi-bin/sprat/public/sprat.pl>

Australian Government. National code of practice for humane shooting of kangaroos and wallabies for commercial purposes. <http://www.environment.gov.au/biodiversity/trade-use/wild-harvest/kangaroo/pubs/code-of-conduct-commercial.pdf>

Australian Government. National code of practice for humane shooting of kangaroos and wallabies for non-commercial purposes. <http://www.environment.gov.au/biodiversity/trade-use/wild-harvest/kangaroo/pubs/code-of-conduct-non-commercial.pdf>

Australian Mammal Society. <http://www.australianmammals.org.au/home.htm>

Australian Wildlife Conservancy. <http://www.actnow.com.au/Groups/Australian_Wildlife_Conservancy.aspx>

IUCN (2010) 2010 IUCN Red List of Threatened Animals. <www.iucn.org/>

Kangaroo Industry Association of Australia. <http://www.kangaroo-industry.asn.au/morinfo/viva.html>

Macropod Program WWF–Australia. <http://www.wwf.org.au>

New South Wales Government, Office of Environment and Heritage. <http://www.environment.nsw.gov.au/wildlifemanagement/KangarooManagementProgram.htm>

Queensland Government. Department of Environment and Resource Management. <http://www.derm.qld.gov.au/wildlife-ecosystems/wildlife/wildlife_permits_and_licences/kangaroo_harvesting.html>

Royal Society for the Protection of Cruelty to Animals, Australia. <http://www.rspca.org.au/>

South Australian Government. Department of Environment and Natural Resources. <http://www.environment.sa.gov.au/Plants_Animals/Abundant_species/Kangaroo_conservation_management>

Western Australian Government. *Management plan for the commercial harvest of kangaroos in Western Australia 2008–2012*. <http://www.environment.gov.au/biodiversity/trade-use/sources/management-plans/pubs/wa-kangaroo-08.pdf>

WWF–The Conservation Organisation. <http://www.panda.org/>

INDEX

www.ingramcontent.com/pod-product-compliance
Lightning Source LLC
LaVergne TN
LVHW060637110826
845147LV00018B/999

9780643097391